T0184099

Lecture Notes of the Institute for Computer Sciences, Social Informatics and Telecommunications Engineering 341

More information about this series at http://www.springer.com/series/8197

Jing Liu · Honghao Gao ·
Yuyu Yin · Zhongqin Bi (Eds.)

Mobile Computing, Applications, and Services

11th EAI International Conference, MobiCASE 2020
Shanghai, China, September 12, 2020
Proceedings

 Springer

Editors
Jing Liu
East China Normal University
Shanghai, China

Honghao Gao
Shanghai University
Shanghai, China

Yuyu Yin
College of Computer
Hangzhou Dianzi University
Hangzhou, China

Zhongqin Bi
Shanghai University of Electric Power
Shanghai, China

ISSN 1867-8211 ISSN 1867-822X (electronic)
Lecture Notes of the Institute for Computer Sciences, Social Informatics
and Telecommunications Engineering
ISBN 978-3-030-64213-6 ISBN 978-3-030-64214-3 (eBook)
https://doi.org/10.1007/978-3-030-64214-3

This Springer imprint is published by the registered company Springer Nature Switzerland AG
The registered company address is: Gewerbestrasse 11, 6330 Cham, Switzerland

Preface

We are delighted to introduce the proceedings of the 11th European Alliance for Innovation (EAI) International Conference on Mobile Computing, Applications and Services (MobiCASE 2020). This conference has brought together researchers, developers, and practitioners from around the world who are interested in mobile computing and leveraging or developing mobile services, mobile applications, and technologies that improve the mobile eco-system.

The technical program of MobiCASE 2020 consisted of 15 full papers which are selected from 49 paper submissions. The conference tracks were: Track 1 – Mobile Application and Framework; Track 2 – Mobile Application with Data Analysis; and Track 3 – AI Application. Aside from the high-quality technical paper presentations, the technical program also featured one keynote speech about service computing and edge computing by Dr. Shuiguang Deng from Zhejiang University, China.

Coordination with the steering chair Imrich Chlamtac and Steering Committee members was essential for the success of the conference. We sincerely appreciate their constant support and guidance. It was also a great pleasure to work with such an excellent Organizing Committee team and we thank them for their hard work in organizing and supporting the conference. In particular, the Technical Program Committee (TPC), led by our TPC co-chairs, Dr. Ying Li, Dr. Yuyu Yin, and Dr. Zhongqin Bi, who completed the peer-review process of technical papers and made a high-quality technical program. We are also grateful to conference manager, Barbora Cintava, for her support and all the authors who submitted their papers to the MobiCASE 2020 conference and workshops.

We strongly believe that the MobiCASE conference series provides a good forum for all researcher, developers, and practitioners to discuss all science and technology aspects that are relevant to mobile computing, applications, and services. We also expect that the future MobiCASE conferences will be as successful and stimulating, as indicated by the contributions presented in this volume.

October 2020

Jing Liu
Honghao Gao
Yuyu Yin
Zhongqin Bi

Organization

Steering Committee

Chair

Imrich Chlamtac   University of Trento, Italy

Members

Ulf Blanke   ETH Zurich, Switzerland
Martin Griss   Carnegie Mellon University, USA
Thomas Phan   Samsung R&D, USA
Petros Zerfos   IBM Research, USA

Organizing Committee

General Chairs

Jing Liu   East China Normal University, China
Honghao Gao   Shanghai University, China

TPC Chair and Co-chairs

Ying Li   Zhejiang University, China
Yuyu Yin   Hangzhou Dianzi University, China
Zhongqin Bi   Shanghai University of Electric Power, China

Local Chair

Weng Wen   Shanghai Polytechnic University, China

Workshops Chair

Yusheng Xu   Xidian University, China

Publicity and Social Media Chair

Jiaqi Wu   Shanghai Polytechnic University, China

Publications Chair

Youhuizi Li   Hangzhou Dianzi University, China

Web Chair

Xiaoxian Yang   Shanghai Polytechnic University, China

Conference Manager

Barbora Cintava   EAI

Technical Program Committee

Gerold, Hoelzl	University of Passau, Germany
Cao, Buqing	Hunan University of Science and Technology, China
Wang, Jian	Wuhan University, China
Wan, Shaohua	Zhongnan University of Economics and Law, China
Chen, Shizhan	Tianjin University, China
Zeng, Jun	Chongqing University, China
Bi, Zhongqin	Shanghai University of Electric Power, China
Xu, Jiuyun	China University of Petroleum, China
Peng, Kai	Huaqiao University, China
Wang, Haiyan	Nanjing University of Posts and Telecommunications, China
Qi, Lianyong	Qufu Normal University, China
Ma, Yutao	Wuhan University, China
Yang, Xiaoxian	Shanghai Polytechnic University, China
Li, Ying	Zhejiang University, China

Contents

AI Application

Mobile Application and Framework

BullyAlert- A Mobile Application for Adaptive Cyberbullying Detection

Rahat Ibn Rafiq$^{(\boxtimes)}$, Richard Han, Qin Lv, and Shivakant Mishra

University of Colorado Boulder, Boulder, CO, USA
rahatibnrafiq@gmail.com,
{richard.han,qin.lv,shivakaht.mishra}@colorado.edu

Abstract. Due to the prevalence and severe consequences of cyberbullying, numerous research works have focused on mining and analyzing social network data to understand cyberbullying behavior and then using the gathered insights to develop accurate classifiers to detect cyberbullying. Some recent works have been proposed to leverage the detection classifiers in a centralized cyberbullying detection system and send notifications to the concerned authority whenever a person is perceived to be victimized. However, two concerns limit the effectiveness of a centralized cyberbullying detection system. First, a centralized detection system gives a uniform severity level of alerts to everyone, even though individual guardians might have different tolerance levels when it comes to what constitutes cyberbullying. Second, the volume of data being generated by old and new social media makes it computationally prohibitive for a centralized cyberbullying detection system to be a viable solution. In this work, we propose BullyAlert, an android mobile application for guardians that allows the computations to be delegated to the hand-held devices. In addition to that, we incorporate an adaptive classification mechanism to accommodate the dynamic tolerance level of guardians when receiving cyberbullying alerts. Finally, we include a preliminary user analysis of guardians and monitored users using the data collected from BullyAlert usage.

Keywords: Cyberbullying · Mobile application · Detection

1 Introduction

Cyberbullying in Online Social Networks (OSNs) has seen an unprecedented rise in recent years. Continued democratization of internet-usage, advancements, and innovations in the area of online social networking and lack of diligent steps to mitigate the effect of cyberbullying by the concerned entities, all have contributed to this unfortunate consequence. The constant threat of cyberbullying in these OSNs has become so expansive and pervasive that it has been reported that in America alone, more than fifty percent of teenage OSNs users have been

J. Liu et al. (Eds.): MobiCASE 2020, LNICST 341, pp. 3–15, 2020.
https://doi.org/10.1007/978-3-030-64214-3_1

affected by the threat of cyberbullying [4]. While real-life bullying may involve verbal and/or physical assault, cyberbullying is different in the sense that it occurs under the umbrella of an electronic context that is available 24/7, thereby rendering the victims vulnerable to its threats on a constant and relentless basis. This unique feature of cyberbullying subjects the victims to devastating psychological effects that later cause nervous breakdowns, low self-esteem, self-harm, clinical depression, and in some extreme cases, suicides [8, 10]. Disturbing events of teens committing suicides after being victimized by cyberbullies have also been reported [9, 30]. Moreover, nine suicide cases have already been attributed to cyberbullying in Ask.fm alone [2]. Cyberbullying has been reported as one of the many potential factors, if not the only factor, for these suicides [3]. Figure 1 shows an example of a cyberbullying instance on Instagram.

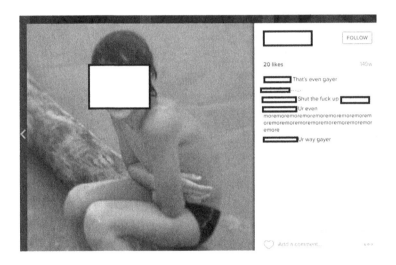

Fig. 1. An example of cyberbullying on Instagram.

There have been numerous works on cyberaggression and cyberbullying by performing thorough analyses of the labeled media sessions from a social network. Cyberaggression is defined as a type of behavior in an electronic context that is meant to intentionally harm another person [16]. Cyberbullying is defined in a stronger and more specific way as an aggressive behavior that is *carried out repeatedly* in OSNs against a person who *cannot easily defend himself or herself,* creating a power imbalance [13, 16, 16, 21, 22, 29]. Developing classifiers for cyberbullying detection for different social networks such as Twitter, Ask.fm, Vine, Instagram, YouTube etc have been proposed as part of most of the previous works [6, 11, 15, 26–28, 35]. System-related challenges such as responsiveness and scalability of a potential cyberbullying detection system have also been proposed [25]. However, there are two key issues that a potential centralized cyberbullying system faces:

- A centralized cyberbullying detection system still needs to have a lot of computing resources to be able to accommodate all the OSNs in the world. Therefore, a system solution that can accommodate all the OSNs currently available will be faced with a daunting challenge of meeting the computational-resource requirements to maintain an acceptable responsiveness
- A potential centralized cyberbullying detection system, such as one presented in [25], makes use of a cyberbullying classifier component that is applied generally to all the guardians. We argue that different guardians will have different tolerance levels, which in turn, might be dependent on their personal preferences, demographic information, location, age, gender, and so on. So, we argue that a system solution that allows different levels of cyberbullying alerts to be sent to parents based on their tolerance levels is the most natural solution.

In this paper, we present the design and implementation of an Android mobile application for guardians: BullyAlert. This mobile application allows the guardians to monitor the online social network activities (currently only supports Instagram) of their kids and get notifications whenever the monitored social network profiles receive a potential cyberbullying instance. The reasons for developing this mobile application are twofold. First, it allows us to delegate classifier computations of cyberbullying detection to the hand-held devices of the guardians, thereby reducing the computational resources needed for a potential centralized cyberbullying detection system. Second, BullyAlert allows the guardians to give the resident classifier feedback about how right or wrong each notification is. The resident classifier then updates itself accordingly to calibrate its tolerance level with that of the guardian using it. This mechanism allows for personalized cyberbullying notification of an individual guardian.

We make the following contributions in this paper.

- We propose the design and implementation of an android application, BullyAlert.
- We present a preliminary user-experience analysis of the guardians who downloaded the mobile application by using the current crop of data
- We present a preliminary comparison the behavior of the users who were being monitored by the guardians with the general population of Instagram to derive some initial key insights by leveraging the current collection of data

2 Related Works

As mentioned earlier, the majority of researches on cyberbullying have focused on improving the accuracy of detection classifiers. Collection, mining and analyses of different social media networks such as Twitter [28], Ask.fm [18], YouTube [5], Instagram [6,12], chat-services [15], Vine [26] have been performed for cyberbullying, cyberaggression [34], harassment [35], aggression under the hood of anonymity [18] and predatory behaviors [14]. As a natural next step, many works have proposed accurate classifiers to detect and predict cyberbullying incidents as well as finding most contributing factors to successful classification of cyber-bullying incidents [6,20].

The scalability and responsiveness challenges of a potential cyberbullying detection system have also been put under the microscope recently [25]. Several systems have been proposed with scalable architectures to detect cyberbullying in social media networks [7,33,35]. However, none of the previous researches addressed the issue of the individual tolerance level of the guardians when it comes to the detection of cyberbullying, which might be a factor dependent on many variables such as the age of the person being monitored, demographics information, gender, and so on. Moreover, due to many novel social networks being introduced every year [32] and the enormous amount of data being generated by the existing ones such as Instagram [17] and YouTube [19], we argue, eventually it will be computationally prohibitive to sustain a central-ized cyberbullying detection architecture. Some mobile applications are available for guardians, however, they are either mostly outgoing packet network sniffing tools [7] or profanity detection apps [31]. None of them incorporates a sophisticated cyberbullying classification or a mechanism to enable dynamic tolerance-level detection.

3   System Design and Implementation

This section presents the design, implementation, and architecture of BullyAlert. We begin by describing the typical user work-flow through a series of use cases, and then present the architecture and implementation of different components of BullyAlert.

3.1   Use Cases

Guardian Registers. After a guardian downloads the application and opens it, the screen in Table 1a is presented. The Guardian has to enter unique email id and password to be able to register into our system. Options to divulge additional information such as age group, gender, and ethnicity, are also provided. Each of these information is presented through a drop-down list. The Guardian can also use the option to decline to give this information.

Guardian Logins. The login screen is shown in Table 1b has two input fields. These fields ask for the email and password used to register into the system. After clicking the log-in button, the guardian is directed to the dashboard of the application.

Guardian Searches for Users to Monitor. The Guardian can search for public Instagram profiles by going to the user search component shown in Table 1c. At first, the guardian selects the social network profile from the drop-down list. *Right now, we are only supporting public Instagram social network*

Table 1. BullyAlert Application

(a) Register   (b) Log In   (c) User Search

(d) User Details   (e) Notification List   (f) Notification Details

profiles. Then in the text field, the username of the user to be monitored is typed. When the search button is clicked, a list of users matching the username entered is shown along with the associated profile pictures for the guardian to facilitate a better identification. To start monitoring a profile, the guardian has to select the profile and then click the monitor button.

Guardian Examines User Profile Information Details. The Guardian can see the basic profile details of the users being monitored, as shown in Table 1d. This page shows guardians the current profile picture, the number of total media shared, and the number of total followers and followings of the user being monitored.

Guardian Gets a List of Notifications. Table 1e shows the screen that the guardian sees when a host of notifications are present in the dashboard. The list has three columns, the first column shows the username of the profile where this cyberbullying notification has originated, the second column indicates the social network and finally, the third column outlines the application's classifier's perceived level of severity for this notification. Currently, the application has two levels of severity, namely low and high. The Guardian can see the details of the notification by clicking the individual notification boxes.

Guardian Examines Notification Details and Give Feedback. To enable the guardians to see the full context of a particular notification and give feedback as to whether the application right or wrong in terms of the severity level, Table 1f is presented. The Guardian has options to click the "see full context" button which will then load not just the latest comments but also the previous comments of that media session. This enables the guardians to get a full picture of the happenings in the media session. The guardian can give feedback through the two buttons, namely right or wrong. This feedback is then used by the application to calibrate its tolerance level according to that of the guardian.

3.2   Architecture and Implementation

This section describes the architecture and implementation of the BullyAlert application's different components. Figure 2 presents the architecture diagram of the BullyAlert system. The guardian communicates with the BullyAlert application for registering, logging in, and getting notifications for potential cyberbullying instances. The application sends guardian data, notification data, and feedback data to the BullyAlert server. The application also contacts the BullyAlert server for authenticating a user log-in. Moreover, the application implements a polling mechanism by which it periodically collects media session data of the Instagram-users (who are being monitored by the guardians) from the Instagram servers.

BullyAlert Server. BullyAlert server is responsible for the following:

– During the registration process, it is responsible for checking that the registration information is verified. It first checks if the email that is being used to register is unique in the system and the password is at least six characters. When the registration is successful, the server stores both the username and the password in an encrypted format. In addition to this, the server stores the

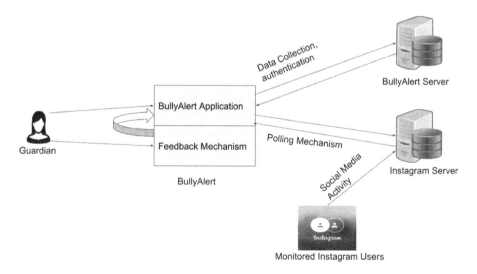

Fig. 2. BullyAlert Architecture

demographic information provided by the guardian through the registration form, such as age group, gender, and ethnicity.
– During login, the server is responsible to check the login credentials of a guardian with the system's stored credentials. It gives an error if the login credentials are not verified which is then in turn shown to the user.
– Storing all the notifications of the guardians. Every time a guardian receives a notification, the application sends the notification meta-data to the server. The data consists of a list of comments for which the notification was raised, the user-name in whose profile the notification was raised, and the severity level of the notification (high or low).
– Storing the feedback that a guardian gives to the application for a particular notification. Every time the guardian gives the application feedback (right or wrong), the application sends the relevant data (list of comments, the application's perceived severity level, the guardian's feedback) to the server.
– Storing each guardian's resident classifier information. This is to facilitate continuity of the guardian's classifier so that when the guardian uninstalls the application, the classifier will keep being stored in our server. This means that, if the guardian, at some later time, chooses to re-install the application, the old classifier will become the classifier of the application instead of the general one.

We have used MongoDB, RESTful API, and node.js for the implementation of this component. The code for this can be found in [24].

Adaptive Classifier. An adaptive classifier for each guardian is more suitable for our application than a general classifier for every guardian because of the potentially subjective nature of cyberbullying. We hypothesize that each guardian will have their tolerance level when it comes to cyberbullying, which in turn, can be dependent on several factors, such as gender, age, race, etc.

The ways we develop this adaptive classifier are as follows. First, we incorporate a feedback mechanism in our application by which, the guardians, upon receiving a potential cyberbullying notification, will be able to give us feedback saying how right or wrong the notification is. *We also show the guardians a list of other media sessions which were not deemed as bullying by our application, to make sure we also get feedback for media sessions which were not in the potential bullying notifications page. This is to enable the application to keep track of the false negatives in addition to false positives.* Second, we use the logistic regression classifier from [25] for the implementation of the application's resident cyberbullying detection component. Every time an instance of the classifier gets feedback, the feedback data encapsulate the media session's list of comments for which the alert was raised and the guardian's label (right or wrong). This datum is considered as a labeled training data for the resident classifier. Feature values described for the logistic regression classifier used in [25] are extracted for each of these feedback sessions. Upon converting the feedback data into training data, we then perform stochastic gradient descent [23] for the resident classifier. Each parent's classifier then reaches a different local optimum, thereby facilitating the adaptive nature of the classifier.

For the guardians whose numbers of feedback are not substantial enough to perform an individual adaptation process, we implement the following. We first collect all the feedback given by all the guardians in our server. Then, based on all this feedback, we update our general classifier that was used by the guardians when they first install our application. We call this *updated general classifier*. This updated general classifier is then propagated to the guardians who don't have enough individual feedback to make sure their classifiers are updated as well. The implementation code can be found in [1].

Polling Mechanism. The polling mechanism is responsible for the following:

– When the guardian searches for a particular user by username, this mechanism fetches the user profiles of which the username-string is a match.
– After a monitoring request of a user by a guardian is approved, the polling mechanism starts polling that user profile every hour for any new posts. This is to make sure the app is updated with the latest media postings of the monitored user.
– In addition to polling for newly posted media, this component is also responsible for getting the newest comments for all the media posted by the user. Every time a host of new comments is posted for a media session, this mechanism fetches those new comments and sends this newest media session data to the adaptive classifier component for classification.

4   User Data Analysis

This section presents a preliminary analysis of the data collected until now from BullyAlert. First, it explores the guardian data and then it performs a comparison of social network behaviors between the general Instagram population [18] and the users who were being monitored by the guardians who downloaded our application.

Table 2. BullyAlert Guardian's Gender and Age Distribution

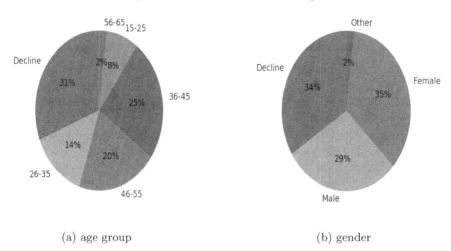

(a) age group   (b) gender

When a guardian registers into our system, in addition to the email and password, we also ask them to provide their gender, ethnicity and age information, if they choose to divulge those. Table 2a and 2b show the distribution of gender and age-group of the 100 guardians who have downloaded BullyAlert until now. From the distributions, it is fairly clear that a substantial portion of the people chose not to provide the demographic information, 31 and 34 percent for age group and gender respectively. In addition to that, the most prominent age group and gender where 36–45 and female respectively.

The reasons we collected this demographic information are twofold. First, when we start to get the classifier feedback data for all the guardians for their different tolerance levels, we will want to investigate if there are any correlations between different guardians' tolerance levels and their demographic information. Second, if we do find that people with the same demographics tend to have the same tolerance levels, we will then want to be able to build different general classifiers for different clusters where each cluster hosts guardians with similar demographics. *While we acknowledge that these 100 guardians' data is an insufficient representation of guardians, we postulate that this preliminary demographics distribution still introduces us to a new systems challenge: what*

Table 3. Comparison between Monitored users of BullyAlert and Instagram population collected in [18]

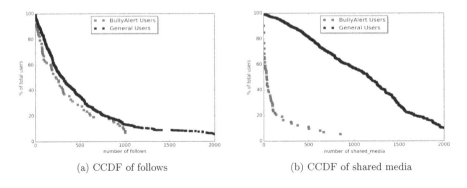

(a) CCDF of follows   (b) CCDF of shared media

about guardians who do not provide demographic information and thus will not belong to any particular cluster by default?

Next, we investigate a comparison analysis between the Instagram users who are being monitored by our application and the general Instagram population, a data-set collected from [18]. First, we compare both sets of users' follow and media-sharing activities. Table 3 and 3b show the CCDF of both set of users' number of people they follow and number of medias they have shared in their profile. It can be seen that the follows activity of both sets follow the same pattern, which is understandable because a user can only follow so many people. But there is a discernible difference in the media sharing activity. The general population's line tends to fall far slower in the graph than that of the BullyAlert-monitored users, with almost 80 percent of the BullyAlert users having less than 100 shared media in their profile. This means that the users who are monitored by the guardians tend to be not as active as the general population. This particular observation also poses an interesting system perspective. Because most of the people who are likely to be monitored by our application will not be sharing as much media, *we can afford to incorporate some sophisticated machine learning classifiers in the application instead of worrying about responsiveness, discussed in* [25]. Again, we like to emphasize here that these are preliminary derivations drawn from our current small set of collected data.

In continuing the narrative, we also put forth a detailed analysis of activities of other people in the user's profile, for example, the likes and comments received in the shared media sessions. Table 4a and 4b show the CCDF of the number of likes and comments received for the media sessions for both set of Instagram users. It can be seen that the media sessions shared by the users being monitored are far less active in terms of getting likes and comments than their general counterparts. *This further solidifies our system perspective that our application's classifier will have fewer data to take care for, thus the classifier does not have to be as lightweight as described in* [25], *based on our current crop of data.*

Table 4. Comparison between Monitored users of BullyAlert and Instagram population collected in [18]

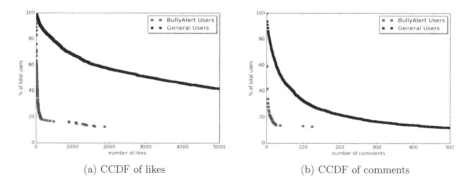

(a) CCDF of likes   (b) CCDF of comments

We acknowledge that the current crop of data is not enough to make a decision, so we plan to keep collecting the data to solidify this preliminary insight.

5   Conclusion

In this paper, we make the following contributions. First, we outline the motivation and design of a mobile application, BullyAlert, that adapts itself according to individual tolerance level for cyberbullying of the guardian. Second, we present a thorough architecture description of the components implemented to develop BullyAlert. Third, we provide a preliminary user analysis of both the guardians and the users being monitored by the application, and in the process, present several potential system issues/ challenges/perspectives using our current crop of data. In the future, we plan to expand our study, collection, and analysis of the guardian data.

References

1. https://github.com/RahatIbnRafiq/AndroidCodesForCyberbullying,   [Online; Accessed 22 October 2018]
2. Broderick, R.: 9 teenage suicides in the last year were linked to cyber-bullying on social network ask.fm. http://www.buzzfeed.com/ryanhatesthis/a-ninth-teenager-since-last-september-has-committed-suicide (2013). [Online; Accessed 14 Jan 2014]
3. Center, C.R.: Cyberbullying research center. http://cyberbullying.us (2013). [Online; Accessed Sep 2013]
4. Council, N.C.P.: Teens and cyberbullying (2007). executive summary of a report on research conducted for National Crime Prevention Council
5. Dadvar, M., Trieschnigg, D., Ordelman, R., de Jong, F.: Improving cyberbullying detection with user context. In: Serdyukov, P. (ed.) Advances in Information Retrieval, pp. 693–696. Springer, Berlin Heidelberg, Berlin, Heidelberg (2013)

6. Dinakar, K., Jones, B., Havasi, C., Lieberman, H., Picard, R.: Common sense reasoning for detection, prevention, and mitigation of cyberbullying (2012). https://doi.org/10.1145/2362394.2362400, http://doi.acm.org/10.1145/2362394.2362400
7. DiProperzio, L.: Cyberbullying applications. http://www.parents.com/kids/safety/internet/best-apps-prevent-cyberbullying/ (2015). [Online; Accessed 6 Feb 2015]
8. Menesini, E., Nocentini, A.: Cyberbullying definition and measurement. some critical considerations. J. Psychol. **217**(4), 320–323 (2009)
9. Goldman, R.: Teens indicted after allegedly taunting girl who hanged herself, bbc news. http://abcnews.go.com/Technology/TheLaw/teens-charged-bullying-mass-girl-kill/story?id=10231357 (2010). [Online; Accessed 14 Jan 2014]
10. Hinduja, S., Patchin, J.W.: Cyberbullying Research Summary, Cyberbullying and Suicide (2010)
11. Hosseinmardi, H., Ghasemianlangroodi, A., Han, R., Lv, Q., Mishra, S.: Towards understanding cyberbullying behavior in a semi-anonymous social network. In: IEEE/ACM International Conference on Advances in Social Networks Analysis and Mining (ASONAM), pp. 244–252. IEEE, Beijing, China (2014)
12. Hosseinmardi, H., Rafiq, R.I., Han, R., Lv, Q., Mishra, S.: Prediction of cyberbullying incidents in a media-based social network. In: Proceedings of the 2016 IEEE/ACM International Conference on Advances in Social Networks Analysis and Mining, IEEE, San Francisco, CA, USA (2016)
13. Hunter, S.C., Boyle, J.M., Warden, D.: Perceptions and correlates of peer-victimization and bullying. British J. Educ. Psychol. **77**(4), 797–810 (2007)
14. Kontostathis, A., West, W., Garron, A., Reynolds, K., Edwards, L.: Identify predators using chatcoder 2.0. In: CLEF (Online Working Notes/Labs/Workshop) (2012)
15. Kontostathis, A.: Chatcoder: Toward the tracking and categorization of internet predators. In: Proceedings of Text Mining Workshop 2009 Held in Conjunction with the Ninth SIAM International Conference on Data Mining (SDM 2009). Sparks, NV, May 2009 (2009)
16. Kowalski, R.M., Limber, S., Limber, S.P., Agatston, P.W.: Cyberbullying: Bullying in the Digital Age. John Wiley & Sons, Reading, MA (2012)
17. Lepage, E.: Instagram statistics. http://blog.hootsuite.com/instagram-statistics-for-business/ (2015). [Online; Accessed 6 Feb 2015]
18. Li, H.H.S., Yang, Z., Lv, Q., Han, R.I.R.R., Mishra, S.: A comparison of common users across instagram and ask.fm to better understand cyberbullying. In: 2014 IEEE Fourth International Conference on Big Data and Cloud Computing, pp. 355–362. IEEE, Sydney, Australia (December 2014). https://doi.org/10.1109/BDCloud.2014.87
19. Mohsin, M.: 10 youtube stats every marketer should know in 2019. https://www.oberlo.com/blog/youtube-statistics (2019). [Online; Accessed 6 Sep 2019]
20. Nahar, V., Unankard, S., Li, X., Pang, C.: Sentiment analysis for effective detection of cyber bullying. In: Sheng, Q.Z., Wang, G., Jensen, C.S., Xu, G. (eds.) APWeb 2012. LNCS, vol. 7235, pp. 767–774. Springer, Heidelberg (2012). https://doi.org/10.1007/978-3-642-29253-8_75
21. Olweus, D.: Bullying at School: What We Know and What We Can Do (1993)
22. Patchin, J.W., Hinduja, S.: An update and synthesis of the research. Cyberbullying Prevention and Response: Expert Perspectives, p. 13 (2012)
23. Python, S.L.: http://scikit-learn.org/stable/modules/sgd.html, [Online; Accessed 22 October 2018]
24. Rafiq, R.I.: https://github.com/RahatIbnRafiq/cybersafetyapp_servercodes, [Online; Accessed 22 October 2018]

25. Rafiq, R.I., Hosseinmardi, H., Han, R., Lv, Q., Mishra, S.: Scalable and timely detection of cyberbullying in online social networks. In: Proceedings of the 33rd Annual ACM Symposium on Applied Computing SAC 2018, pp. 1738–1747. ACM, New York, NY, USA (2018). https://doi.org/10.1145/3167132.3167317, http://doi.acm.org/10.1145/3167132.3167317

26. Rafiq, R.I., Hosseinmardi, H., Han, R., Lv, Q., Mishra, S., Mattson, S.A.: Careful what you share in six seconds: detecting cyberbullying instances in vine. In: Proceedings of the 2015 IEEE/ACM International Conference on Advances in Social Networks Analysis and Mining 2015, pp. 617–622. ACM, Paris, France (2015)

27. Rajadesingan, A., Zafarani, R., Liu, H.: Sarcasm detection on twitter: a behavioral modeling approach. In: Proceedings of the Eighth ACM International Conference on Web Search and Data Mining WSDM 2015, pp. 97–106. ACM, New York, NY, USA (2015). https://doi.org/10.1145/2684822.2685316, http://doi.acm.org/10.1145/2684822.2685316

28. Sanchez, H., Kumar, S.: Twitter bullying detection. In: NSDI, p. 15. USENIX Association, Berkeley, CA, USA (2012)

29. Smith, P.K., del Barrio, C., Tokunaga, R.: Principles of Cyberbullying Research. Definitions, measures and methodology, Chapter: Definitions of Bullying and Cyberbullying: How Useful Are the Terms? Routledge (2012)

30. Smith-Spark, L.: Hanna smith suicide fuels calls for action on ask.fm cyberbullying, CNN. http://www.cnn.com/2013/08/07/world/europe/uk-social-media-bullying/ (2013). [Online; Accessed 14 Jan 2014]

31. Thom, B.: SafeChat: Using Open Source Software to Protect Minors from Internet Predation. Ursinus College (2011). https://books.google.com/books?id=SLbQvQEACAAJ

32. Ueland, S.: 10 new social networks for 2019. https://www.practicalecommerce.com/10-new-social-networks-for-2019 (2019). [Online; Accessed 6 Sep 2019]

33. Villatoro-Tello, E., Jurez-Gonzlez, A., Escalante, H.J., Gmez, M.M.Y., Pineda, L.V.: A two-step approach for effective detection of misbehaving users in chats. In: Forner, P., Karlgren, J., Womser-Hacker, C. (eds.) CLEF (Online Working Notes/Labs/Workshop). CEUR Workshop Proceedings, vol. 1178. CEUR-WS.org (2012)

34. Willard, N.: Cyberbullying and cyberthreats: Responding to the challenge of online social aggression, threats, and distress. Research, Champaign, IL (2007)

35. Yin, D., Xue, Z., Hong, L., Davison, B.D., Kontostathis, A., Edwards, L.: Detection of harassment on web 2.0. In: Proceedings of the Content Analysis in the WEB 2, 1–7 (2009)

An Optimization of Memory Usage Based on the Android Low Memory Management Mechanisms

Linlin Xin[1], Hongjie Fan[2(✉)], and Zhiyi Ma[2]

[1] Advanced Institute of Information Technology, Peking University, Beijing, China
1501220090@pku.edu.cn
[2] School of Electronics Engineering and Computer Science, Peking University, Beijing, China
{hjfan,mazhiyi}@pku.edu.cn

Abstract. When users manipulate low memory Android devices, they frequently encounter the application problem of loading slowly because of limited amount of memory. In particular, more applications installed, problems will occur more frequently. We deeply observe the low memory management mechanism of the Android system and find the system has some shortcomings, such as memory recovery efficiency, unnecessary memory requests. In this paper, we optimize memory usage by improving recovery efficiency, prioritize the use of less memory, prevent the instantaneous increase in memory usage, and reduce unnecessary memory requests. Experimental results in a real environment show that our methods effectively increase the size of free memory, and reduce the phenomenon of application self-startup and association startup.

Keywords: Performance optimization · Low memory management · Auto startup

1 Introduction

Android is a Linux-based, open source operating system which runs on smartphones, tablets, smart TVs, and smart wearable devices [1]. Compared to other embedded systems, Android system has good open source features. Programmers can quickly develop applications without compromise application compatibility, and IT vendors can easily provide feature-specific devices to meet diverse and complex needs.

However, as the number of loading applications increases, the memory requirements make it impossible for running some applications smoothly on low-memory devices. It causes users to load system slowly, especially when more applications installed. Ultimately, the user's experience will be affected under this situation.

As shown in Fig. 1, The effect of system memory is demonstrated in the Google I/O Conference[1]. With the process goes from the top to bottom, the importance is weakened in turn. Basic functions are affected or may even the system is restarted when the system terminates processes such as Home, Service, Perceptible, and Foreground.

---

[1] https://events.google.com/io2018/.

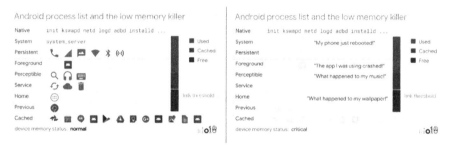

Fig. 1. System memory impact on users.

This phenomenon mainly due to the lack of system memory by two reasons:

1. Android is a multi-tasking system based on Linux system [1]. The corresponding application is executed according to the priority and queue order. Under the situation of low system memory and high CPU load, tasks such as user operations are to be queued for execution and the system is unresponsive.
2. In order to reclaim a portion of the memory, the system has to clean up most of the file cache or even terminate some of the application process. When these applications are used again, the system needs to reallocate the memory to load the file resources, which resulting in a large time expenditure.

As the number of applications increases, the system free memory becomes less. In order to ensure the system has a certain amount of free memory, the system's process will reclaim part of the file cache. Android Low Memory Killer (LMK)[2] is a process monitoring memory and reacting to high memory pressure by killing the least essential process(es) to keep system performing at acceptable levels. Low Memory Killer (LMK) is important whenever available memory of system is below some threshold values [2]. The Low Memory Killer (LMK) is an android specific implementation of OOM Killer (Out Of Memory Killer)[3] mechanism based on Linux. It sets a few adjustment value and minimum free memory pairs while booting OS. LMK is triggered if the amount of available memory is not sufficient. Then LMK kills apps which having a lower adjustment value if the amount of available memory is under a certain threshold. In Android platform, these thresholds are set using min-free values whereas in Linux kernel using watermark levels [3, 4]. Kook et al. [5] proposed a novel selection scheme which runs the OOM killer to terminate arbitrary processes in O(1) time.

The LMK selection process satisfies the following two conditions:

1. Oom_score_adj. Oom_score_adj is calculated by LMK through oom_adj and needs to be greater than the preset min_score_adj value.
2. Largest amount of physical memory. After the process is selected, the LMK sends a SIGKILL signal that cannot be ignored or blocked to terminate the process.

---

[2] https://android.googlesource.com/platform/system/core/+/master/lmkd/README.md.

[3] https://en.wikipedia.org/wiki/Out_of_memory.

For example, the memory size is configured to 2 GB machines for Android 7.1 system, the low memory threshold rules are shown in Table 1.

Table 1. Rule table of low memory killer in low memory threshold

Free memory is lower than	72 MB	90 MB	108 MB	126 MB	216 MB	315 MB
Corresponding min_score_adj value	0	58	117	176	529	1000
Corresponding oom_adj value	0	1	2	3	9	16

When the system memory is small than 315 MB, LMK would select the appropriate process with oom_adj value high than 16 and terminates it. The retaining process then greatly reduce the startup time when the user is used again. However, in some special cases, LMK has two shortcomings since the low memory threshold and the oom_adj correspondence value are fixed:

First, when the free memory is lower than 216 MB but higher than 126 MB, LMK would terminate the process with the oom_adj value larger than or equal to 9 until the free memory is higher than 216 MB. However, if the system whose oom_adj value is greater than or equal to 9 is all terminated and the free memory is still less than 216 MB, the system will scan all processes again to find a process with an oom_adj value greater than or equal to 9. At this point, most of the time is wasted on the scanning operation, which makes the memory recovery inefficient.

Second, LMK is a passive memory recycling mechanism triggered by the memory threshold level, which consumes a lot of CPU resources. The conditions for terminating the process are only by two factors: the oom_adj value of the process and the size of physical memory occupied by the process. Because the selection process standard is simple, and sometimes even terminates the application that user cares about. In addition, some useless processes are not preferentially reclaimed because they do not reach the preset low memory threshold. For example, when the free memory is just below 216 MB, according to Table 1, the system can terminate the process with the oom_adj value greater than or equal to 9, but there are some inactive background services (B Service processes with oom_adj value of 8). It is not recovered because it is below the preset threshold of 9.

Based on these observations, we deeply study the low memory management mechanism and optimize the system based on shortcomings. We optimize the two aspects of preventing the instantaneous increase of memory usage and reducing the unnecessary memory application. In summary, contributions in this paper are as follows:

1. Modify the Linux kernel layer to improve the recovery efficiency of Low Memory Killer. We introduce the vmpressure formula to measure the memory recovery pressure of the current system. When the system pressure is high, the system selectively terminates the process with a smaller oom_adj value, which helps LMK reduce the number of scans and improve the efficiency of memory recovery.
2. Modify the activity manager service part of application framework layer to preferentially recycle less memory. For service processes that are always working in

the background (for example, B Service), we increase oom_adj value so that LMK can preferentially terminate the least recently used process to free up more memory space.

3. Modify the active service in the application framework layer to prevent the instantaneous growth of memory usage. According to a large number of test results, we adjust the number of parallel startups. Thereby we can reduce the instantaneous memory usage and prevent the pulsed downward trend of the free memory size.

4. Modify the application layer to reduce unnecessary memory requests. We restrict application self-start and association startup for services at application and system level, reducing unnecessary memory requests.

The first two parts are corresponding to supplement and the LMK deficiency optimization. If the system is in a low memory state after optimizing the LMK, we will optimize the latter two parts to avoid the system being in a low memory state.

The rest of the paper is organized as follows: In Sect. 2, we present the related studies and compare those with our study. We describe our platform architecture and methodology in Sect. 3. In Sect. 4, we present our performance evaluations and discussions. Finally, we present the conclusions, and future work in Sect. 5.

2   Related Work

Purkayastha [6] proposed a survey of Android optimization about Android architecture and Android runtime. Hui [7] presents a novel schema for high performance Android systems with using pre-cache on multiple mobile platform. Mario [8] performed an in-depth analysis into whether implementing micro-optimizations can help reduce memory usage. Besides, Mario conducted a survey with 389 open-source developers to understand how they use micro-optimizations to improve the performance of Android apps. Joohyun [9] develop a context-aware application scheduling framework which adaptively unloads and preloads background applications for a joint optimization saving and minimizing the user discomfort from the scheduling.

Android OS [10] has a process terminating function, called Low Memory Killer. The function automatically terminates application processes when the size of available memory become small. There are many approaches proposed to optimize the low memory killer based on app launch patterns. The method proposed in [11] follows the standard framework in Android low memory killer. Different from the standard killer, the approach redefines the importance hierarchy using different options, e.g., the LRU heuristic, app re-forking time, or the highest important level within the two, etc., and then reprioritizes the apps inside a same importance level with metrics derived from the memory consumption of the apps. Without a systematic way, those options are combined in an ad-hoc manner, leading to the experimental result that app re-forking time and derived metrics from app memory consumption do not help on top of the importance level redefined with the LRU heuristic. Their best result is achieved with an option similar to the 2nd baseline in our experiment. Cong [12] proposed the approach to optimize the low memory killer with reinforcement learning. The low memory killer continuously observing various indicators and metrics for memory management, making the process-killing decisions, and taking app launch latencies as the penalties from the decision-making environment. The low memory killer function automatically terminates application processes when the size of available memory becomes small. Sang-Hoon et al. [13] proposed a novel memory reclamation scheme called *Smart LMK*, which minimizes the impact of the process-level reclamation on user experience. The memory footprint and impending memory demand are estimated from the history of the memory usage keeping an app. *Smart LMK* picks up the least valuable apps and efficiently distinguishes the valuable apps among cached apps and keeps those valuable apps in memory.

Kim et al. [14] proposed heuristic approach to detect the periodic patterns and show that it improves the performance on some specific apps. Researchers in [15] have proposed a complex Markov decision model for reclamation of memory on Android. They periodically inspect the stop queue to calculate the survival probability of app in the next inspection and those having low probabilities are killed based on a threshold. Zhang et al. [16] presents an approach for Android devices which protects certain processes from memory acquisition by process memory relocation. They relocated to the special memory area where the kernel is loaded. Yu et al. [17] propose a two-level software rejuvenation, with the two levels referring to software applications and the OS. Based on this strategy, they construct a Markov regenerative process model to optimize the time required to trigger rejuvenation for Android smartphones. The methods are complementary to ours in terms of refining our simple non-parametric models to predict app launches by embedding the contexts (e.g., the location context) as the model conditions. Some researches monitor how users use the apps and evaluate the interestingness of an app with linear combination of different features [18]. Liang et al. [19] analyze that current memory management algorithms are not working well on Android smartphones and exploiting the tradeoff. Amalfitano et al. [20] present FunesDroid for the automatic detection of memory leaks tied to the Activity Lifecycle in Android apps to detect and characterize these memory leaks. Lee et al. [21] design an estimator to minimize the Frobenius norm of the gain matrices which show excellent performance in environments, where noise information is unknown and in which sudden disturbances are inserted. Ryusuke et al. [22] focus on Android's Generational GC and propose a method for improving its promotion condition. They control promotion based on monitored statistical information that indicates smaller objects tend to die in shorter time. Qing et al. [23] propose MEG, a Memory-efficient Garbled circuit evaluation mechanism, which utilizes batch data transmission and multi-threading to reduce memory consumption.

For improving the real-time capabilities, without loss of original Android functionality and compatibility to existing applications, Igor et al. [24] apply the RT_PREEMPT patch to the Linux kernel, modify essential Android components like the Dalvik virtual machine and introduce a new real-time interface for Android developers. Wook et al. [25] propose a personalized optimization framework for the Android platform which takes advantage of user's application usage patterns in optimizing the performance. Based on this, they implement an app-launching experience optimization technique which tries to minimize user-perceived delays and state loss when a user launches apps. Instead of refining the policy of process-killing, there are other efforts to improve the app launch performance with system level optimization. Some authors attempted to improve performance by modifying AMS. Yang et al. [26] changed the LRU based replacement policy in the AMS. They suggested a pattern-based replacement algorithm. Ju et al. propose the reclaiming method for mitigating sluggish response based on the MOS Kernel module [27], besides before each app launch demonstrating that the number of kernel function calls is reduced, which make the sluggish response when launching applications. Ahn [28] proposed a system that automatically detects and corrects memory leaks caused by JNI. The system works in detection, correction, and verification. Yang et al. [29] developed an automated tool to acquire the entire content of main memory from a range of Android smartphones and smartwatches. Kassan et al. [30] presents a new self-management of energy based on Proportional Integral Derivative controller (PID) to tune the energy harvesting and Microprocessor Controller Unit (MCU) to control the sensor modes. Maiti et al. [31] present a framework enabling effective and flexible smartphone energy management by cleanly separating energy control mechanisms from management policies.

3   Platform Architecture and Methodology

The overall structure is shown in Fig. 2. First, we use Vmpressure to calculate the memory recovery pressure of the current system by the ratio of Scanned Page Size and Recovered Page Size. According to Vmpressure, we estimate current memory recovery pressure. We terminate a qualifying process and reclaim memory if the system is in low memory. In addition, we provide the priority to recover less used memory called Activity Manager Service. The system updates the oom_adj value of the process according to the importance. If oom_score_adj larger than min_score_adj, we will select this process and turn into the terminate process. Additionally, we reduce the memory usage rate by increasing the maximum number of concurrent starts of the service to alleviate the pressure of memory recovery and give the CPU enough time to recycle. Besides we reduce unnecessary memory requests from the perspective of restricting application self-starting and association startup.

3.1   Improve the Memory Recovery Efficiency

In order to improve the memory recovery efficiency of LMK, we introduce the vmpressure formula to measure the memory recovery pressure of the current system. The main idea of the vmpressure formula is to use the ratio of the unsuccessfully reclaimed memory size to the scanned Page size as a measure of the system memory recovery pressure. The vmpressure formula calculates the current instantaneous memory recovery pressure. The larger the ratio, the system has the greater memory recovery pressure. So when the recovery pressure is large, the process is terminated (the min_score_adj value is lowered). It is worth mentioned that the initial instantaneous pressure index is not very accurate. But over time, this ratio will gradually be averaged and refined to accurately represent the memory recovery pressure of the current system. When the system pressure is large, the system will terminate the process with small oom_adj value and improve the efficiency of memory recovery.

The formula is shown in:

$$Pressure = \frac{Mem_{Scanned} - Mem_{Reclaimed}}{Mem_{Scanned}}$$

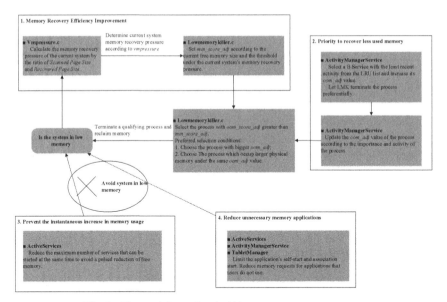

Fig. 2. The workflow of android low memory management.

Where Pressure is the pressure value of the memory recovery, the variable Mem$_{Scanned}$ is the scanned memory size, and the variable Mem$_{Reclaimed}$ is the size of the reclaimable memory. The formula shows that the more memory that is not successfully reclaimed, the memory has the greater recovery pressure of the system.

After the pressure value Pressure is calculated, the reclaimed memory path determines whether to decrease the min_oom_adj value according to the size of the Pressure value and the size of the file cache. The specific rules are shown in Table 2:

Table 2. Optimized low memory killer low memory threshold override rule table.

Free memory is lower than	72 MB	90 MB	108 MB	126 MB	216 MB	315 MB
Original min_score_adj	0	58	117	176	529	1000
Corresponding oom_adj	0	1	2	3	9	16
After the leap, min_score_adj	0	58	58	58	176	176
The corresponding oom_adj after the leap	0	1	1	1	3	3

1. When the pressure value pressure $>= 95$, and file cache is $<= 250$ MB, the process is terminated.
2. When the pressure value is $95 >$ pressure $>= 90$, and file cache is less than 175 MB, the process is terminated.

3. If the above two conditions are not met, it is considered that the memory recovery pressure is not large, the memory is reclaimed according to the original threshold.

3.2 Prioritize the Use of Less Memory

Low Memory Killer is a passive memory recycling mechanism triggered by the memory threshold level, so it takes up a lot of CPU resources when triggered [12]. The system terminates the process in two conditions: the oom_adj value of the process and the size of physical memory occupied, which are analyzed previously.

In order to prevent the useless process from being reclaimed due to failure to reach the preset low memory threshold, we prevent the system from being in a lower memory state by preferentially recycling fewer processes. The main idea of this method is according to a large number of test verifications, we find the service process working in the background (the B Service process in the code, the oom_adj value is 8), and select the least recently used process according to the order in the LUR list. Since when the process is terminated, only one oom_adj value of the least recently used B Service process is incremented each time, the next B Service process is selected, so that the design is to reclaim less memory while still being used. Keep the process as much as possible. Thus, Low Memory Killer preferentially terminates the least recently used process to free up more memory.

3.3 Prevent the Instantaneous Increase in Memory Usage

The Android system is a multitasking system that allows multiple services to start at the same time. However, when the number of concurrent services is large, the free memory will be pulsed down and resulted a sudden shortage of free memory. As a result, the system will be busy with reclaiming memory work, causing the work of other processes to be delayed. After in-depth research, many applications are secretly launched by the service component without the user's knowledge. Starting the application creates the application's process, the creation process requests a block of memory. Therefore, when the number of concurrent services in the service is large, the instantaneous memory usage will increase, causing a sudden shortage of memory space in the system. In this situation, we reduce the memory usage rate by increasing the maximum number of concurrent starts of the service to alleviate the pressure of memory recovery and give the CPU enough time to recycle.

The phenomenon of service parallel startup is more common when the system sends a system broadcast. For example, after booting, the system will issue a system broadcast with startup completion, such as BOOT_COMPLETED. All event will be started, which will start a service to deal some operations in the background. The default number of service parallel launches for Android system is 8. After extensive testing and analysis, we change the number to 4. If the value is too small, there will be too many services that are queued to start, resulting in a long event delay and affecting the application of system-level services. For example, the alarm application may cause the alarm time to expire, but the alarm service is still in the queue, which brings a bad user experience.

3.4   Reduce Unnecessary Memory Requests

Some applications are used to reduce the first boot time or start by the user without knowing by service or broadcast. It wastes the system memory resources and affects the speed of the system operation. Taking the Lenovo YOGA Table 3 X90F tablet as an example, as shown in Fig. 3, although a new machine does nothing after booting (no application is installed, no network is connected), the current memory state is captured as below after 10 min.

Table 3. Detailed configuration parameters of YOGA flat 3.

Model	YOGA Table 3 X90F
Operating system	Android 6.0.1
System memory	2 GB
Storage	32 GB
CPU model	Intel 2.24 GHz(Quad-Core)
Screen size	10.1 inches
Screen resolution	1280 * 800
Screen ratio	16:10
WiFi function	Support

```
eileen@eileen-VirtualBox:~$ adb shell cat /proc/meminfo
MemTotal:        1945188 kB
MemFree:          130824 kB
MemAvailable:     615220 kB
Buffers:           23976 kB
Cached:           838512 kB
SwapCached:          280 kB
Active:           728700 kB
Inactive:         770820 kB
```

Fig. 3. Memory status after the device is turned on for 10 min.

From the figure, you can find the device with 2 GB memory, minus some memory reserved by the kernel. The actual physical memory (*Mem Total*) of the device is 1945188 KB, about 1.85 GB. The free memory (*Mem Available*) after 10 min of booting is 615220 KB (about 600 MB), and the memory space is already low. If we install some applications that start automatically, the device will be in low memory state after booting, and the system will be busy reclaiming memory. Therefore, it is necessary to reduce unnecessary memory requests. From the perspective of restricting application self-starting and association startup, the system sends two nodes, Broadcast and Start Service, to optimize.

3.4.1 Broadcast Process Optimization

The broadcast is sent to the corresponding *Broadcast Receiver* through the *Activity Manager Service*. First, the *Activity Manager Service* will save all the registered *Broadcast Receiver* to the *Receiver Resolver* object, and parse the incoming data, and find the corresponding *Broadcast Receiver* according to the value of the Intent. Then create a new *Broadcast Record* block with the obtained parameters and add it to the *Broadcast Queue*. The *Activity Manager Service* maintains two *Broadcast Queues* (the foreground *Broadcast Queue* d broadcast queue and the background broadcast queue), which hold all the Broadcast objects that need to be sent. Finally, the *process Next Broadcast* method of the is called by the message mechanism to sequentially process the broadcast in the queue. The transmission and processing is asynchronous.

We add restrictions on sending broadcasts in the *process Next Broadcast* () method of *Broadcast Queue* so that broadcasts that meet the restrictions are not sent. These restrictions correctly intercept broadcast events that are not user-operated, non-system events. After in-depth analysis and extensive testing of Android code, this article adds the following restrictions to the transmission of Broadcast.

1. Broadcast can send if Broadcast belongs to the user-initiated application (that is, the application of the process in the LRU list).
2. Broadcast can be sent if the Intent action of Broadcast is started as a Widget.
3. Broadcast can be sent if the package name of the broadcast is the same as the caller's package name.
4. In other cases, the system checks the policy set by the user and decides whether it can be sent according to the policy.

The above is a restriction policy for sending broadcasts, but most applications are self-starting by receiving broadcasts of Android system events, such as system boot broadcast, device networking broadcast, caller broadcast. These system broadcasts are an integral part of the normal operation of the system. It is no longer possible to make a detailed distinction between the applications here, so it is necessary to make policy restrictions on the startup.

3.4.2 Optimize System Services

Service in the Android system is divided into two categories according to the level: application services and system services. Application services are services defined and implemented in an application. These services are managed in *Active Services*. System services refer to the services necessary for the Android system to work, such as *Activity Manager Service, Package Manager Service*, etc., such services are managed in the *Service Manager*. There is a clear difference between the two. Since the middle layer has been standardized in Android design, when the application developer implements a service, it is only necessary to simply implement the server and the proxy. Since this article is to limit the background self-starting of the application, it is only for the Service startup process in *Active Service*. In order to facilitate developers, the Android system encapsulates the Service. So the developer only needs to call the *start Service* or *bind Service* method of the Context. After the call, *Context Impl* will call the *Active Services* retrieve *Service Locked* method through the corresponding method of *Activity Manager Service* to get the Service record.

The calling process is shown in Fig. 4.

Fig. 4. System service call flow.

Since the two startup methods of Service finally obtain the Service record through the *retrieve Service Locked* method, this article makes policy restrictions when the *retrieve Service Locked* method obtains the Service record. After in-depth analysis and extensive testing of the Android code, the following restrictions are imposed on the startup of the Service.

1. When the *Service Info* of the Service is null, it can be started.
2. When the Service belongs to a user-initiated application (that is, the application of the process in the LRU list), it can be started.
3. When calling the service's process's uid < 10000 and the Service's Intent Action is *Sync Adapter*, check the policy set by the user, and decide whether it can be started according to the policy.
4. When calling the service's process's uid < 10000 and the permission is BIND_JOB_SERVICE, check the policy set by the user and decide whether it can be started according to the policy.
5. When the process of the service is uid < 10000, it can be started, and only the service of the three-party application is restricted.
6. When the uid of the process where the service is located is the same as the uid of the caller, it can be started.
7. When the application package name of the Service is the same as the caller's package name, it can be started.
8. In the remaining cases, check the policy set by the user and decide whether it can be started according to the policy.

The above strategy design is mainly used to limit the self-starting and association startup of the application. Its structure can be divided into the following four parts:

1. Call the interface of the *Tablet Master Service* in the Android system framework. For example, in the *Re-Service Service Locked* method of *Active Services*, the interface of the *Tablet Master Service* is called to determine whether the Service can be started.
2. Launch the interface part of the management to provide the user with an operable interface setting. Users can set which applications can be self-starting and which applications can be launched by association to set up a whitelist.
3. Store the user-set application self-start, association start, and whitelist data into the backend database.

4. The preset configuration file is used to save the list of applications that the user cannot control. There is some system preset applications that do not want to be restricted by users, such as self-starting of applications.

These four parts work together to help users limit the self-starting and association startup of the application. The working flow chart is shown in Fig. 5

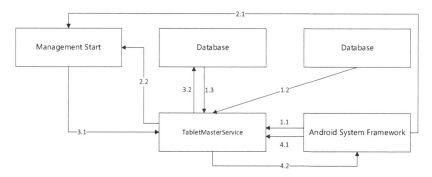

Fig. 5. Workflow chart limiting application self-starting and association startup.

The process in Fig. 5 indicates that when the device is powered on, the *Service Manager* of the Android system starts the *Tablet Master Service*. Then perform the 1.2 operation, 1.2 indicates that the *Tablet Master Service* reads the preset list file, and sets the policy value of the applications that are not controllable by these users to 0. (*Tablet Master Service* divides the application into four categories according to the policy value. The first type of policy value is 0, which is an application that is not controllable by the user, that is, an application that does not participate in the restriction condition. The second policy value is 1, which is set by the user in the application interface. The application is not allowed to start. The third policy value is 2, which is the application that the user is set to allow to start in the application interface. The fourth policy value is 3, which is the application that the user sets to the whitelist in the application interface. Execute startup restrictions for applications in the whitelist). Then perform the 1.3 operation, 1.3 means that the *Tablet Master Service* reads the user settings from the database, and saves the application's package name and policy value in the form of *Hash Map* key-value pairs, and transmits them to the application interface part through the interface.

When the user clicks on the app, the user is listed with a list of all apps that can be restricted, and a switch button is provided for the user to personalize. When the user clicks on the application, the 2.1 operation is performed, and 2.1 indicates that the application obtains a list of all installed applications from the system's *Package Manager*. Then, the 2.2 operation is performed. 2.2 indicates that the application obtains the policy values of the self-starting and association startup of each application from the interface in the *Tablet Master Service*, and merges the list with the 2.1 part into a list, and correctly displays the status of each application on the interface.

When the user modifies the management policy value through the interface, the 3.1 operation is performed. 3.1 indicates that the application invokes the *Tablet Master Service* interface to modify the corresponding policy value. Then perform the 3.2 operation, 3.2 indicates that the *Tablet Master Service* saves the policy value.

When the system sends a broadcast or starts the service, the 4.1 operation is performed. 4.1 indicates that the *process Next Broadcast* () method of the *Broadcast Queue* calls the *check Intent Auto Start* interface of the *Table Master Service* to determine whether the broadcast satisfies the restriction condition and can be sent. Or the *refresh Service Locked*() method of *Active Services* calls the *check Service Auto Start* interface of the *Tablet Master Service* to determine whether the Service satisfies the restriction condition. 4.2 Operation indicates that the *Tablet Master Service* notifies the system of the result, and the system sends or does not send Broadcast, start or not start the Service according to the result.

The above is the workflow of the application. Each module has a clear division of labor and work together to effectively reduce the self-starting and associated startup of the application.

4   Results and Discussion

4.1   Experimental Setup

We run different apps on a mobile device with Intel quad-core, 2 GB memory for performance test and result verification. Detailed parameter configuration information of device is shown in Table 3.

The version of the device was fairly smooth before being optimized. However, after installing 30 mainstream applications, system is often jamming during use. We prepare two YOGA Table 3 X90F tablet devices with the same hardware parameters, one for the native Lenovo system ROM: YT3_X90F_10_row_wifi_20180202, labeled "T-original"; the other is the optimized ROM, labeled "T-optimized". 80 applications are installed on the two tablets. The application list is shown in Fig. 6.

From the perspective of user operation, the system response speed is the main criterion for measuring system performance, such as the time when the application interface switches display, the time when the application is launched after clicking the application icon. From a system internal point of view, memory usage is the primary measure of system performance.

We compare the data of two devices before and after optimization to test whether the solution can optimize system performance and optimization in three ways:

1. Calculate the state of device free memory.
2. Filter analysis and statistics intercepts.
3. Count the interface response speed.

58tongcheng-8.1.1.apk	douyin-1.7.3.apk	MoxiuLauncher-6.3.16.apk	TencentComic-7.11.6.apk
360mobilesafe-7.7.4.apk	douyutv-3.6.1.apk	NetEaseMusic-4.3.4.apk	TencentKaraoke-4.5.5.275.apk
360video-4.3.8.apk	Faceu-3.0.1.020318.apk	pinduoduo-3.56.0.apk	TencentSecure-7.6.0.apk
12306-3.0.1.01221000.apk	huluxia.gametool-3.5.1.74.1.apk	pingan.lifeinsurance-4.12.0.apk	TencentVideo-5.9.2.13908.apk
Alipay-10.1.15.463.apk	hunanTV-5.6.4.apk	PPtv-7.2.5.apk	thunder-5.54.2.5330.apk
Amap-8.2.8.2146.apk	huyazhibo-5.6.3.apk	qiyivideo-9.0.0.apk	tudou-6.16.3.apk
B612camera-7.0.4.apk	iFlyIME-8.0.6367.apk	QQ-7.3.8.apk	UCbrowser-11.8.6.966.apk
baidu.netdisk-8.2.0.apk	iphoneassess-7.0.4.apk	QQbrowser-8.2.0.3950.apk	ugc.live-3.3.0.apk
baidu-10.3.0.11.apk	jingdong-6.6.4.apk	QQlite-3.6.2.apk	unicom-5.6.2.apk
baiduhomework-9.10.0.apk	jinritoutiao-6.5.9.apk	QQmusic-8.0.1.5.apk	unionpay-5.0.5.apk
BaoFeng-7.5.02.apk	kankan-2.9.4.apk	QQpim-6.8.0.apk	vstudy-3.20.0.apk
BeautyCamera-7.3.60.apk	kuaikan.comic-5.0.0.apk	QQreader-6.5.9.888.apk	WatermelonVideo-2.3.9.apk
bilibili-5.22.1.apk	kugouplayer-8.9.4.apk	renrenshipin-3.6.6.apk	WeChat-6.6.2.apk
CCB-4.0.8.apk	kuwoplayer-8.6.1.0.apk	ringtoneduoduo-8.6.0.1.apk	Weibo-8.1.2.apk
ChinaABCBank-3.7.3.apk	Kwai-5.5.3.5776.apk	shuqi-10.6.5.60.apk	WIFIlookpassword-3.1.5.apk
chinatelecompay-6.6.0.apk	le123video-2.4.4.apk	sogouinput-8.17.apk	xfplay-5.0.0.0.apk
cleanmaster-6.03.5.apk	letv-7.9.2.apk	sohuvideo-6.9.1.apk	xianyu-6.0.3.apk
CMCC-4.3.0.apk	meituan.takeoutnew-6.2.3.apk	TableGame-1.9.8.1.apk	xiaoyuansouti-6.11.0.apk
Connotations-6.8.0.apk	meituan-9.0.1.apk	taobao-7.5.1.apk	XimalayaFM-6.3.69.3.apk
Didi-5.1.32.apk	meituxiuxiu-7.2.0.0.apk	tencent.ttpic-5.4.5.1828.apk	youku-7.1.1.apk

Fig. 6. Application list.

4.2   Experimental Results

4.2.1   System Free Memory

By monitoring the free memory, first we test the solution to improve memory recovery efficiency and prioritize the use of less memory methods to avoid long-term low memory. We use the free memory in the device process as a measure to improve memory reclamation efficiency and prioritize the use of less memory. During the test, we connect the device to the computer using the USB cable, and execute the following command on the computer to check the current memory status of the system:

We conduct two devices (T-original device and T-optimized device) as follows:

(1)   After powering on, let it stand for 10 min. Then we check the current memory status of the system, and record the value of "*Mem Available*" (which is the size of free memory).

(2)   Open the app list and tap the app's icon. After the system displays the application interface, record the value of "*Mem Available*".

(3)   Press the Home button to return to the desktop.

(4)   According to the application list, we cycle step 2 and 3 and use the "*Mem Available*" value after the first 15 applications to compare the results of the two devices before and after optimization.

Results are shown in Table 4.

Table 4. Comparison of average free memory.

Operation	Free memory on "T-original" device (Before Optimization)	Free memory on "T-optimized" device (After Optimization)
Power on and leave for 10 minutes	363 MB	615 MB
After using "Connotations"	343 MB	519 MB
After using "Watermelon Video"	237 MB	482 MB
After using "China Mobile"	274 MB	393 MB
After using "Alipay"	229 MB	383 MB
After using "YouKu"	210 MB	314 MB
After using "iFly IME"	207 MB	309 MB
After using "Xiao Yuan Sou Ti"	211 MB	273 MB
After using "Xian Yu"	128 MB	270 MB
After using "XimalayaFM"	125 MB	484 MB
After using "WeChat"	161 MB	387 MB
After using "Weibo"	199 MB (Application Not Responding)	343 MB
After using "NetEase Music"	67 MB (System Crash)	313 MB
After using "Table Game"	222 MB	342 MB
After using "Tencent Video"	183 MB	397 MB
After using "Tencent Comic"	149 MB	338 MB
The Average of Free Memory	**207 MB**	**385 MB**

As shown in Table 4, the memory of T-original of power-on reset is 363 MB after 10 min, which is slightly higher than the operating threshold of 315 MB for Low Memory Killer. After using 2 applications, the free memory is lower than Low Memory Killer's working threshold. As the number of applications increases, the memory resources are in short supply, and the minimum is 67 MB, which may cause crash in the system process. The T-optimized average free memory is 385 MB, and the lowest memory is 270 MB. It is still within the primary threshold range of the Low Memory Killer. At this time, the memory recovery pressure is not too large.

In addition, we view the memory usage of the system by setting the select memory page. After the device is turned on after standing for 24 h, we can analyze the statistics within 3 h of the memory page.

(1) By monitoring the free memory, test whether can improve the memory recovery efficiency. Results are shown in Fig. 7:

We observe the device before optimization for 10 min after the boot is still low, 363 MB, slightly higher than the LMK operating threshold of 315 MB after use 2 applications. The free memory is lower than the working threshold of LMK. As the number of applications used increases, the memory resources are in short supply and the minimum is 67 MB, which causes the system to restart. The optimized device has an average free memory of 385 MB and a minimum memory of 270 MB, which is still in the primary threshold of LMK setting.

(2) By monitoring the memory consumption of the device boot process (because when booting the device needs to start more services, so we need to optimize the number

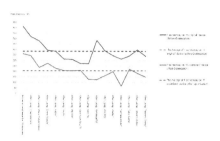

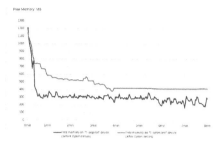

Fig. 7. Comparison of free memory before and after optimization.

Fig. 8. Variation of free memory during boot process.

of parallel boot of the boot process service), we test prevention of instantaneous growth of memory usage. Results are shown in Fig. 8.

We observe that before optimization ("T-original" device) within 0.5 min of booting, the free memory dropped from about 1200 MB to 300 MB, indicating that the device's instantaneous memory usage is very high. The optimized device ("T-optimized" device) slowly dropped from around 1150 MB to 400 MB within 4 min of booting. It proves that the instantaneous growth of memory usage prevention effectively avoids the pulse growth of device memory and alleviate system CPU pressure.

4.2.2 Self-start and Association Startup Application

In order to eliminate the impact of reducing the unnecessary memory application on the optimization, we set the limit of the T-optimized device from the startup and the associated startup to no limit. Then conduct the following experiments for the two devices ("T-original" device and "T-optimized" device):

(1)   Restart the device and execute the./meminfo.sh script.
(2)   Stand for 8 min after booting up and use the script to count the *"Mem Available"* value within 8 min to compare the results of the two devices before and after optimization.

Through the monitoring system log, reducing unnecessary memory requests can effectively intercept application self-starting and association startup. By counting the number of logs without any operation status within 30 min after booting, we show the effectiveness. Results are shown in Table 5.

As shown in Table 5, before optimization ("T-original" device), 52 processes are started by the *Broadcast Receiver* component without any operation, and the cumulative number of startups is 273. 31 processes have been restarted repeatedly. 50 processes are started by the Service component. The cumulative number of startups is 404. Among them, 36 processes have been restarted repeatedly. The most frequently restarted is the "Google GMS" framework, up to 94 times. After analysis, it is found that these processes are repeatedly restarted, causing the device to be in a low memory state. So

Table 5. Process data started by broadcast before and after optimization.

	Before optimization	After optimization
Proc. for broadcast	273 times (52 processes started. 31 of them were restarted, and the most frequent restarts were Google Search applications, up to 31 times.)	50 times (41 processes started, only 8 of them started repeatedly, The number of restarts is up to 3 times.)
Proc. for service	404 times (50 processes are started. 36 of them are restarted repeatedly, and the most frequently restarted are Google GMS applications, up to 94 times.)	9 times (9 processes started, there is no process to restart)

Low Memory Killer needs to continuously terminate the process to release the space, and the terminated process will start again. The optimized device ("T-optimized" device) has 41 processes that are started by the Broadcast Receiver component within 30 min of booting. The cumulative number of startups is only 50, only 8 processes are restarted. The most recent reboot was the "Google Play" app, which was only restarted 3 times. 9 processes are started by the service component and there is no restarted process. The specific situation is shown in Fig. 9 and Fig. 10

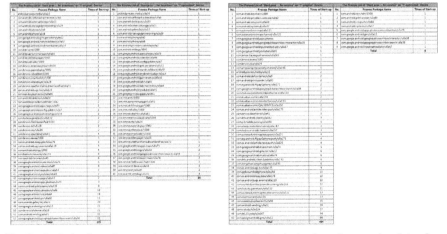

Fig. 9. Optimization of processing data before and after the broadcast started.

Fig. 10. Optimization of processing data of pre/post service startup.

The proposed method of reducing unnecessary memory can effectively reduce unnecessary process startup and reduce unnecessary memory overhead.

4.2.3 Response Time

We short the response time of response and improve system performance by counting the time difference between operation and interface display. During the experiment, we recorded video for each of two devices in the following operations:

(1) Leave it on for 10 min after turning it on;
(2) Open the application list and click the app's icon to record the time point when the user pressed the icon as the start time.
(3) After the system displays the application interface, record the time point of the display interface as the end time.
(4) According to the application sequence in the application list, cycle step 2 and 3. We respectively count the click time point and interface display time point of the top 15 applications. Then we calculate the length of time the system responds to the user operation. Results are shown in Fig. 11.

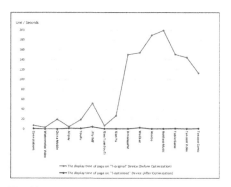

No.	APP	"T-original" Device (Before Optimization)			"T-optimized" Device (After Optimization)		
		Click time	Display time	The display time of Page (Unit / Second)	Click time	Display time	The display time of Page (Unit / Second)
1	Connotations	00:00:04.05	00:00:10.22	6.561	00:00:16.22	00:00:17.24	1.066
2	Watermelon Video	00:00:21.18	00:00:24.21	3.095	00:00:32.20	00:00:32.29	0.297
3	China Mobile	00:01:27.01	00:01:46.19	19.528	00:00:40.26	00:00:41.06	0.033
4	Alipay	00:02:05.04	00:02:07.28	2.792	00:00:55.04	00:00:57.24	2.066
5	YouKu	00:02:21.05	00:02:38.21	17.561	00:01:08.16	00:01:09.27	1.163
6	iFly JME	00:03:20.18	00:04:08.00	47.396	00:01:21.04	00:01:25.18	4.462
7	Xiao Yuan Sou Ti	00:04:25.12	00:04:31.24	6.396	00:01:31.34	00:01:32.08	0.462
8	Nan Yu	00:05:43.15	00:06:10.01	26.528	00:01:48.22	00:01:49.11	0.627
9	XimalayaFM	00:07:15.10	00:09:45.21	150.368	00:03:13.04	00:03:13.17	0.429
10	WeChat	00:11:44.27	00:14:16.14	151.561	00:03:31.13	00:03:34.13	3
11	Weibo	00:17:16.03	ANR 00:20:25.06	189.098	00:04:43.07	00:04:44.02	0.825
12	NetEase Music	00:21:56.08	System Crash	195	00:05:53.29	00:05:53.23	0.792
13	Table Game	00:02:18.09	00:04:48.13	150.132	00:04:17.26	00:04:18.27	1.031
14	Tencent Video	00:05:32.23	00:07:55.11	144.066	00:05:09.21	00:05:10.12	0.693
15	Tencent Comic	00:09:56.26	00:11:48.26	112	00:05:30.01	00:05:20.25	0.934
	The average time			81.7			1.2

Fig. 11. Time statistics of T-original and T-optimized.

Fig. 12. Time comparison chart of T-original and T-optimized.

For ease of observation, as shown in Fig. 12.

It can be found from the test results that before optimization ("T-original" device), response for a long time. Especially in the case of 7 applications after startup, the display interface takes 2–4 min and even the application does not respond (Weibo) or system crash (Neteasy Music) and other serious problems. The optimized device ("T-optimized" device) has an average response time of 1.2 s. The response time of only 2 applications is longer, around 3–4 s, and no application is unresponsive or system crash.

By increasing the size of the free memory and reducing the application self-starting or association startup, the system can effectively shorten the response time of the user. In addition, it is also found that limiting the application self-starting or association startup significantly reduce the number of broadcasts and services.

Based on the above verification, we demonstrate the efficiency of low memory recovery mechanism and avoid the device being in a low memory state for a long time.

4.3 Discussion

It is worth mentioned that this solution is optimized for the lack of low memory management mechanism, which is different from the third-party application for optimizing mobile phone memory, such as 360 security guard. At present, the three-party application on the market mainly provides two types of functions: optimizing the memory and limiting the application self-starting. The former has three shortcomings:

1. They terminate all application processes, including those are in use. Our method terminates an application that the user does not want to close, and the application reload when the application is recently used.
2. The memory can be cleaned only when the user uses the one-click function to clear the memory. Therefore, real-time monitoring and optimization cannot be achieved, and the effect is not durable. Although some applications provide timing cleanup, which lead to frequent reloads of the switch application. The program optimizes memory from within the system, with higher authority and obvious effects. And the process will be terminated only when the system is in a low memory state, which has less impact on the user.
3. It cannot be customized to the product, and is not the best choice. We adjust the low memory threshold appropriately according to the memory parameters of the product. Our solution is optimized from the inside of system, so it can be customized.

5 Conclusions

Mobile applications usually can only access limited amount of memory, especially in low memory situation. In this paper we analyze the principle of the Low Memory Killer mechanism and point out the reasons for the low recovery efficiency and error recovery process of LMK. After that, we propose the optimization scheme for avoiding the low memory state. Experimental results show that our methods effectively increase the size of free memory, reduce the phenomenon of application self-startup and association startup. In the future, we plan to further extend our approach suitable for more applications. Besides we will consider new combined methods or caching strategies to optimize memory usage.

Acknowledgments. This work is supported by the National Natural Science Foundation of China (No. 61672046).

References

1. Annuzzi, J., Darcey, L., Conder, S.: Introduction to Android Application Development: Android Essentials. Addison-Wesley Professional (2015)
2. Nomura, S., Nakamura, Y., Sakamoto, H., Hamanaka, S., Yamaguchi, S.: Improving choice of processes to terminate in Android OS. GCCE, pp. 624–625 (2014)
3. Gorman, M.: Understanding the Linux Virtual Memory Manager. Prentice Hall Professional Technical Reference (2004)

4. Mauerer, W.: Professional Linux Kernel Architecture. Wiley Publishing, Inc. Technical Reference (2008)
5. Joongjin, K., et al.: Optimization of out of memory killer for embedded Linux environments. In: Proceedings of the 2011 ACM Symposium on Applied Computing. ACM (2011)
6. Purkayastha, D.S., Singhla, N.: Android optimization: a survey. Int. J. Comput. Sci. Mob. Comput.-A Mon. J. Comput. Sci. Inform. Technol. **2**(6), 46–52 (2013)
7. Zhao, H., Chen, M., Qiu, M., Gai, K., Liu, M.: A novel pre-cache schema for high performance Android system. Future Gener. Comp. Syst. **56**, 766–772 (2016)
8. Vásquez, M.L., Vendome, C., Tufano, M., Poshyvanyk, D.: How developers micro-optimize Android apps. J. Syst. Softw. **130**, 1–23 (2017)
9. Lee, J., Lee, K., Jeong, E., Jo, J., Shroff, N.B.: CAS: context-aware background application scheduling in interactive mobile systems. IEEE J. Sel. Areas Commun. **35**(5), 1013–1029 (2017)
10. Nagata, K., Yamaguchi, S., Ogawa, H.: A Power Saving Method with Consideration of Performance in Android Terminals. UIC/ATC, pp. 578–585 (2012)
11. Nomura, S., Nakamura, Y., Sakamoto, H., Hamanaka, S., Yamaguchi, S.: Improving choice of processes to terminate in Android OS. GCCE 2014, pp. 624–625 (2014)
12. Li, C., Bao, J., Wang, H.: Optimizing low memory killers for mobile devices using reinforcement learning. In: 13th International Wireless Communications and Mobile Computing Conference (IWCMC), pp. 2169–2174 (2017)
13. Kim, S.-H., Jeong, J., Kim, J.-S., Maeng, S.: SmartLMK: a memory reclamation scheme for improving user-perceived app launch time. ACM Trans. Embedded Comput. Syst. **15**(3), 47:1–47:25 (2016)
14. Kim, J.H., et al. A novel android memory management policy focused on periodic habits of a user. Ubiquitous Computing Application and Wireless Sensor, pp. 143–149. Springer, Dordrecht (2015)
15. Yang, C.-Z., Chi, B.-S.: Design of an Intelligent Memory Reclamation Service on Android. TAAI 2013, pp. 97–102 (2013)
16. Zhang, X., Tan, Y., Zhang, C., Xue, Y., Li, Y., Zheng, J.: A code protection scheme by process memory relocation for android devices. Multimedia Tools Appl. **77**(9), 11137–11157 (2017). https://doi.org/10.1007/s11042-017-5363-9
17. Yu, Q., et al.: Two-level rejuvenation for android smartphones and its optimization. IEEE Trans. Reliab. (2018). https://doi.org/10.1016/j.ress.2017.05.019
18. Kumar, V., Trivedi, A.: memory management scheme for enhancing performance of applications on Android. In: 2015 IEEE Recent Advances in Intelligent Computational Systems (RAICS). IEEE (2015)
19. Liang, Y., Li, Q., Xue, C.J.: Mismatched Memory Management of Android Smartphones. HotStorage (2019)
20. Amalfitano, D., Riccio, V., Tramontana, P., Fasolino, A.R.: Do memories haunt you? An automated black box testing approach for detecting memory leaks in android apps. IEEE Access **8**, 12217–12231 (2020)
21. Lee, S.S., Lee, D.H., Lee, D.K., Kang, H.H., Ahn, C.A.: A Novel Mobile Robot Localization Method via Finite Memory Filtering Based on Refined Measurement. SMC 2019, pp. 45–50 (2019)
22. Ryusuke, M., Yamaguchi, S., Oguchi, M.: Memory consumption saving by optimization of promotion condition of generational GC in android. In: 2017 IEEE 6th Global Conference on Consumer Electronics (GCCE). IEEE (2017)
23. Yang, Q., Peng, G., Gasti, P., Balagani, K.S., Li, Y., Zhou, G.: MEG: memory and energy efficient garbled circuit evaluation on smartphones. IEEE Trans. Inform. Forensics Secur. **14**(4), 913–922 (2019)

24. Kalkov, I., Franke, D., Schommer, J.F., Kowalewski, S.: A Real-Time Extension to the Android Platform. JTRES 2012, pp. 105–114 (2012)
25. Song, W., Kim, Y., Kim, H., Lim, J., Kim, J.: Personalized optimization for android smartphones. ACM Trans. Embedded Comput. Syst. **13**(2 s), 60:1–60:25 (2014)
26. Yang, C.-Z., Chi, B.-S.: Design of an Intelligent Memory Reclamation Service on Android. TAAI 2013, pp. 97–102 (2013)
27. Ju, M., Kim, H., Kang, M., Kim, S.: Efficient memory reclaiming for mitigating sluggish response in mobile devices. ICCE-Berlin 2015, pp. 232–236 (2015)
28. Ahn, S.: Automation of Memory Leak Detection and Correction on Android JNI. MobiSys 2019, pp. 533–534 (2019)
29. Yang, S.J., Choi, J.H., Kim, K.B., Bhatia, R., Saltaformaggio, B., Xu, D.: Live acquisition of main memory data from Android smartphones and smartwatches. Digital Invest. **23**, 50–62 (2017)
30. Kassan, S., Gaber, J., Lorenz, P.: Autonomous energy management system achieving piezoelectric energy harvesting in wireless sensors. Mob. Netw. Appl. **25**(2), 794–805 (2019). https://doi.org/10.1007/s11036-019-01303-w
31. Maiti, A., Chen, Y., Challen, G.: Jouler: A Policy Framework Enabling Effective and Flexible Smartphone Energy Management. MobiCASE 2015, pp. 161–180 (2015)

Design of a Security Service Orchestration Framework for NFV

Hu Song[1], Qianjun Wu[2], Yuhang Chen[2], Meiya Dong[3(✉)], and Rong Wang[4]

[1] State Grid Jiangsu Electric Power Co., Ltd. Information and Communication Branch, Nanjing, China
[2] Information System Integration Company, NARI Group Corporation, Nanjing, China
[3] Taiyuan University of Technology, Taiyuan, China
dongmeiya@163.com
[4] China Energy Investment Corporation, Beijing, China

Abstract. Network functions virtualization (NFV) emerges as a promising network architecture that separates network functions from proprietary devices. NFV lowers the cost of hardware components and enables fast and flexible deployment of network services. Despite these advantages, NFV introduces new security challenges. Currently, there is little research on a holistic framework to solve these security issues. In this paper, we propose a security service orchestration framework that can construct a cooperative working mechanism for NFV security. We present the demand analysis and describe the system design principles and implementation details. We propose a lightweight holistic architecture design of the security service orchestration system to solve current security issues. The system's effectiveness is also shown based on technical review.

Keywords: NFV · Network security

1 Introduction

NFV (Network Functions Virtualization) [11] is an emerging technology that decouples network functions from dedicated devices and thus provides an agile and cost-efficient way to deploy network functions and services. The advantages of NFV have been recognized by both industry and academia [7,15,16]. Since it can reduce the cost of ownership by implementing network services on commercial off-the-shelf hardware, telecommunication giants such as AT&T are embracing NFV architecture and providing business solutions based-on NFV [1,2]. According to a recent report [3], the global NFV market size is projected to reach 36.3 billion USD by 2024.

However, the successful deployment of NFV inevitably introduces new challenges in network security management. For example, since NFV turns network functions into software modules, an attack on a software module might affect

© ICST Institute for Computer Sciences, Social Informatics and Telecommunications Engineering 2020
Published by Springer Nature Switzerland AG 2020. All Rights Reserved
J. Liu et al. (Eds.): MobiCASE 2020, LNICST 341, pp. 37–52, 2020.
https://doi.org/10.1007/978-3-030-64214-3_3

others located on the same virtual machine. In addition, many NFV systems are built on open-source projects such as Openstack and OSM. Potential software bugs in these projects might lead to security threats [4].

In recent years, much efforts have been made to improving NFV security [8,9,18,20]. Basile et al. [5] proposed a software component named Policy Manager, which allows users to specify their security requirements and automatically selects required virtual network functions based on policy refinement techniques. In [6], the authors further added support for automatic enforcement of security policies as a part of the Orchestrator, and designed an optimization model which was able to select the best way to refine network policies. Marchetto et al. [13] proposed a network modeling approach for formal verification of forwarding behavior. Pedone et al. [17] designed a security framework to protect end-users which is compliant to the Security-as-a-Service paradigm. While all these studies focus on solving NFV security issues, a holistic framework is still in need.

To this end, we aim to design software-defined security architecture and provide a systematic framework to ensure security in NFV in this paper. It can reasonably allocate virtual security device resources and deliver various security services to improve adequate security protection to achieve intelligent security services. The rest of this paper is organized as follows. In Sect. 2, the overall architecture and requirements analysis of NFV are discussed. In Sect. 3, the architecture design of the security service orchestration system is analyzed, and an overall NFV security service orchestration architecture is proposed. Section 4 concludes the full paper.

2   NFV Architecture

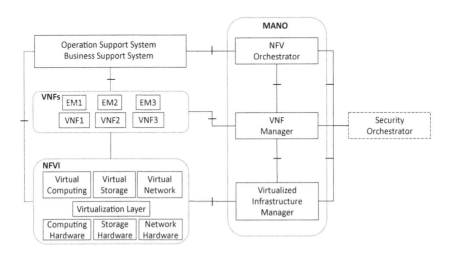

Fig. 1. NFV architecture

As shown in Fig. 1, an NFV architecture consists of Network Functions Virtualization Infrastructure (NFVI), Virtualized Network Functions (VNFs), and Management and Network Orchestration (MANO).

- NFVI is built on the basis of non-proprietary hardware like x86 servers. It abstracts computing hardware, storage hardware, and network hardware into virtual resources, and provisions these resources to support the executions of VNFs.
- VNFs are virtualization of network functions. They are deployed and running on virtual machines. Element Management Systems (EMs) are responsible for fault, configuration, accounting, performance, and security monitoring VNFs.
- MANO are composed of Virtualized Infrastructure Manager (VIM), VNF Manager, and NFV Orchestrator (NFVO). VIM controls and manages computing, network, and storage resources. VNF Manager is responsible for the lifecycle management of VNF instances. And NFVO performs global resource management across multiple VIMs and maintains network services of VNFs.
- Operation Support System (OSS) and Business Support System (BSS) are a collection of system management applications, supporting network management, configuration management, customer management, etc.

In this paper, we also propose a Security Orchestrator as part of the framework. It can both connect these functions and make real-time security assessment on the process of data flow of the whole NFV system.

2.1 Demand Analysis

The design of security orchestration system for NFV should include the following components:

(1) Web business layer: The front-end display of the orchestration system is the entrance to the interaction between the orchestration system and the user, and it is also the generator of the orchestration strategy. The Web layer can generate corresponding orchestration strategies according to the actual needs of users, and deliver them to the orchestration engine.
(2) Application development standards: In the orchestration system, security devices of different manufacturers are accessed through a proxy App. The agent App northbound interacts with the orchestration engine by following the "standard", while the southbound converts the "standard" into device interface data to achieve the adaptation of the device to the orchestration system. Developers develop applications under standards. Generally speaking, the development standard can be either a software development SDK or an API interface specification. This system uses the API interface specification.
(3) Orchestration strategy: The orchestration system needs to provide a set of description specifications of the orchestration logic, which is convenient to transform the orchestration scenario into a program parseable orchestration strategy.

Table 1. Demand analysis

Functional requirement	Explain
Web Business Layer	Web services for orchestration system based on App Store project development
Layout engine	Parse the security policy under the Web, and realize the task generation, scheduling and execution
Safety Controller	equipment management module, resource scheduling module, service chain module
Virtual security equipment management platform	Virtual security device startup, registration, network configuration and device information query interface
Application development standards	Develop development standards for proxy applications of security equipment for vendor adaptatio
Organization strategy	Provides a description mechanism to translate choreographed scenarios into standard security policies

(4) Orchestration engine: The orchestration engine is the core module of the orchestration system. Its main functions include orchestration strategy analysis, orchestration task scheduling, and execution.

(5) Orchestration scheduling module: The orchestration scheduling module is located inside the security controller to implement the scheduling function of security resources. In order to improve the effective utilization of safety resources, the centralized scheduling of safety resources needs to adopt a suitable resource scheduling algorithm. The input of the resource scheduling algorithm includes various security policy requirements issued by the application plane on the one hand, and the security resource information obtained from the data plane on the other. The security policy of the application plane can reflect the security needs of users, and the data of the data plane can reflect the actual situation of security resources. If the requirements of the application plane and the security resource capabilities of the data plane can be reasonably matched using the resource scheduling algorithm, it can improve the utilization rate of security equipment, reduce the queuing time, and speed up the system's response to security threats.

(6) Virtual security device management platform: It effectively manages various virtual security devices under a virtual network environment. In the security service orchestration system, it is responsible for the startup and configuration of the underlying security device and provides the device status query interface and device configuration interface to the security controller (Table 1).

3   Security Service Choreography System Architecture Design

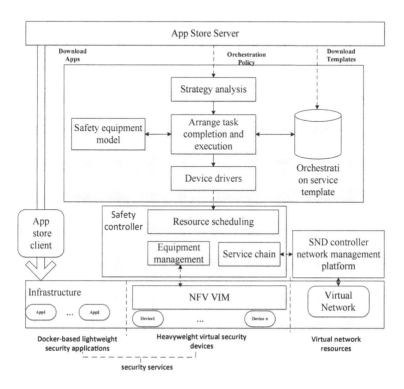

Fig. 2. Overall framework of security service orchestration system architecture

Based on previous analysis of the requirements for the security service orchestration system, we divide the orchestration system into four major parts for design and development, namely the web business layer, the orchestration engine, the relevant modules within the security controller, and the virtual security device management platform. Among them, the relevant internal modules of security control include equipment management module, a resource scheduling module, and a service chain module. This topic implements the service chain module in general scenarios (Fig. 2).

The top layer of the architecture is a web business layer developed based on the AppStore platform, which is generally deployed on a central server to provide users with web interaction services and generate orchestration strategies. Below the Web layer is the client environment. The deployed modules include orchestration engines, security controllers, and infrastructure layers and services. The orchestration engine is responsible for receiving orchestration policies and orchestration service templates issued by the AppStore, generating orchestration tasks, and scheduling execution. The security controller receives the resource invocation request of the orchestration engine and selects appropriate equipment from the resource pool to perform the protection task through the resource scheduling module. The infrastructure layer provides the computing, storage, and network resources required for security services to operate. The service chain module of the security controller is interfaced with the SDN control platform to realize the secure service chain and provide a guarantee for the correct implementation of the orchestration strategy at the bottom.

3.1 System Workflow

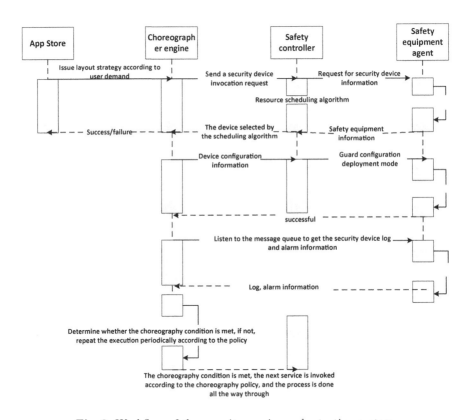

Fig. 3. Workflow of the security service orchestration system

Firstly, AppStore recommends related security orchestration scenarios to users based on their asset information. After the user selects and customizes the orchestration method of the security service, the AppStore sends the corresponding orchestration strategy to the orchestration engine. Then, the orchestration engine analyzes the orchestration strategy and generates a Job for it (persistent task). Finally, the orchestration engine executes according to the scheduling method specified by the policy. The orchestration strategy supports flexible job execution methods, such as executing once every 5 min, keeping the task all the time, or stopping the task after executing it several times (Fig. 3). The following details the job execution process:

(1) In the beginning, the orchestration engine calls the first security application of the orchestration scene according to the orchestration strategy. If the application itself can complete the security service, the security task execution result is returned directly to the orchestration engine.
(2) If the security service requires a security device to implement, the security application sends a device call request to the security controller. Request parameters include protection target information, security device type, and device configuration parameters.
(3) The safety controller selects the appropriate safety equipment through the resource scheduling algorithm according to the safety equipment type and other business parameters and sends a protection task to the selected device.
(4) The safety controller collects the task execution results of the safety equipment and returns the information to the safety application. The security application further processes the collected logs and alarm information, and converts it into an interface format recognized by the orchestration engine, then returns it to the orchestration engine.
(5) Finally, the orchestration engine compares the information returned by the security application with the orchestration strategy. If the triggering condition is met, the next security application is called for protection according to the orchestration strategy. If the triggering condition is not met, the first one is called again according to the task scheduling method defined by the orchestration policy security application.

3.2 AppStore Layer

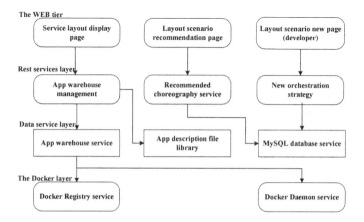

Fig. 4. AppStore cloud architecture

The web business architecture of the orchestration system is shown in Fig. 4. The AppStore layer is divided into four sublayers, from bottom to top: Docker sublayer, data service sublayer, REST service sublayer, and WEB sublayer. Among them, the Docker sublayer is used as the underlying operating environment. The AppStore platform uses Docker container technology to package the App and provides services such as Docker image management and container management. The data service sublayer implements data persistence and saves the orchestration strategy information that has been issued. The REST service sublayer implements the encapsulation of business logic and provides REST API to invoke. The WEB sublayer provides interface-related UI for smooth user operation.

AppStore is the entrance to various network security services. Currently, it is mainly a web security scan. After selecting the corresponding security service, it fills in the relevant parameters to issue the security service. This implementation case is a web vulnerability scan. In the security controller, several main modules used are App Managers, which are used to manage the supported applications. Device Manager manages various information of the currently enabled devices, such as IP address ports and service types. EventManager is the control of internal interactive events, and the multiple modules interact through subscription push events. BootAgent controls the startup and recycling of virtual machines, and the resource pool is the server resource cluster where web vulnerabilities are located.

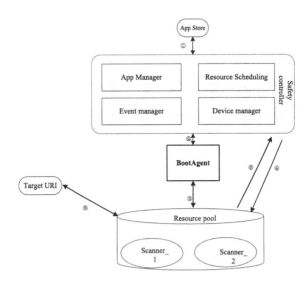

Fig. 5. Resource pool

In Fig. 5, the infrastructure resource pool of the architecture can generally reach the scale of thousands of servers. With the introduction of virtualization technology, the size of server nodes (VMs) will further increase; meanwhile, the cloud computing resource pool improves resource utilization. The data traffic generated by server resources of the same size is bound to be more potent than traditional construction methods.

Data Center Network Hierarchical Model. With the introduction of NFV, the physical resources at the infrastructure layer are pooled into virtual resources to provide software functions for the business layer. A data center based on the NFV architecture should have sufficient bandwidth resources and various network facilities. In order to ensure the normality of virtual network elements. In operation, the internal management mechanism seems particularly relevant. The designed hierarchical model of the data center network is shown in Fig. 6.

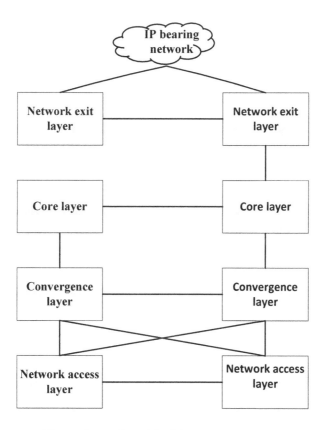

Fig. 6. Layered model of data center network

Data centers should ensure the flexible division of network areas. They are constructing an NFV-based data center that meets the data center design principles. Different application systems can be divided into different areas.

There must be a core layer that is responsible for core switching functions. It is the core part of the entire network and acts as a transmission bus. Compared with other levels, the features of the system are enhanced by directly adding equipment at the core switching layer to improve scalability and availability. At the same time, the access device must be connected to the network access layer to complete the function of the device accessing the network. An aggregation layer is added between the network access layer and the core layer to complete the aggregation function of the access device. Finally, externally accessed resources need to increase the network exit layer, and at the same time, add network firewalls, load balancing, and other equipment to the data center inside and outside. This layered design can not only ensure the security of each region, increase flexibility, but also have good scalability.

In order to ensure that the network elements deployed in the data center can communicate securely and follow the principles of flexibility, reliability, and scalability, the network is divided into network exit layers, core layers, aggregation layers, and network access layers using horizontal partitioning and vertical layering Four levels, as shown in Fig. 7:

(1) The network exit layer is used to connect the internal network and the external network. At the same time, it can play the function of internal and external network information conversion and control the information forwarding of the internal and external network.
(2) The core layer is connected to the network exit layer and connected to the convergence layer, which mainly implements the core switching equipment information forwarding and control.
(3) The aggregation layer is connected to the core layer and connected to the aggregation layer, which mainly implements the aggregation function of the access layer equipment.
(4) The service access layer mainly provides network access functions to ensure regular access to terminal equipment. Considering the high reliability, high availability, and security of the network, the deployment plan is as follows: Configure VLAN + VRF on the core switch to achieve isolation from other business systems and configure QOS to meet RCS (converged communication) service bandwidth requirements.

In the server's internal virtualization layer, the virtual switch (VSW) function should be supported to meet the requirements of throughput, CPU, and memory utilization, to achieve virtual machine switching, VLAN differentiation, and port speed limit functions.

3.3 Orchestration Engine Design

The top layer of the Orchestration Engine is the REST API module, which provides orchestration policy registration/deregistration interface, registered orchestration policy query interface, and Job query interface to the AppStore platform. The middle layer is the core function module of the orchestration engine, which implements the orchestration strategy analysis and orchestration task scheduling functions. Among them, the security device model is a high abstraction of security devices by the orchestration engine, and each type of security device corresponds to a device model in the engine. The orchestration service template defines the mapping relationship between the input and output of two apps, which can be dynamically loaded by the orchestration engine and is the critical design to break the barriers between applications. The lowest level is the database module and the security service driver module. The database module is used to store the registered orchestration strategies and job information to achieve data persistence. The security service driver is used to connect the orchestration engine with different manufacturers and different types of security devices. It is independent of the orchestration engine in the form of App (agent application). The agent application is developed in the north direction according to the SDS equipment standard API interface specification and receives the security protection tasks issued by the orchestration engine. The south direction interfaces with the security controller and is used to issue protection dispatch information for equipment dispatch requests and equipment identification (Fig. 8).

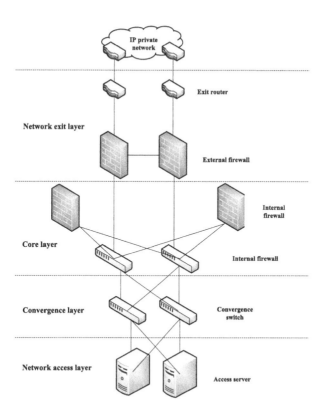

Fig. 7. Data center network layering scheme

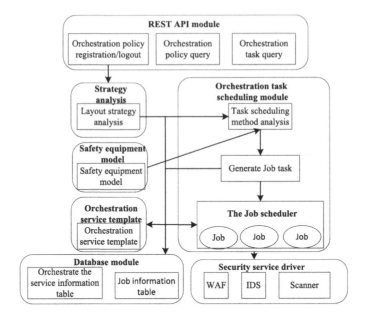

Fig. 8. Orchestration Engine Design Framework

3.4 Virtual Security Management Platform

The virtual security device management platform is written in Python and combines open source technologies such as DNSmasq, libvirt, OpenvSwitch, and Linux namespace. In this section, we explain the solution of this system from

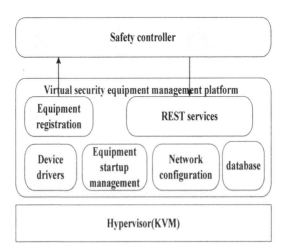

Fig. 9. Virtual Security Management Platform

the perspective of device startup and registration, device network management, and database selection (Fig. 9).

(1) Device startup and registration

In the software-defined security architecture, the security resource pool must shield the differences between different manufacturers and different types of devices, and at the bottom layer, it must implement unified management of device startup and registration. Since security devices are all virtualized in the security resource pool, the management of security devices is equivalent to the control of virtual machines. Establish a virtual security device model that is as general as possible. If a device is different from the general model, the user can implement personalized functions in subclasses by inheriting the general class. In other words, the management platform's access to new equipment is implemented by adding "plug-in classes." In addition, when the management platform is started, it reads the configuration file information, starts a number of security devices in advance, and uniformly registers with the security controller. The configuration file contains the type, number, image information, and drive information of the boot security device.

(2) Equipment network management

The equipment management platform network topology design is shown in Fig. 10. In the picture, WAF (Web Application Firewall) and RASS (Remote System Evaluation System) are virtual security devices, each with three network cards, which are the management port, data stream entrance, and data stream exit. Correspondingly, these three network cards are respectively connected to three OVS (OpenvSwitch) bridges of br-con, br-in, and br-out. Among them, the management port is used to transmit management layer data, and the data flow in and out is used to transmit network traffic. The above two security device deployment modes are also common deployment modes of the management platform. Since the device does not have a business request when it is initially started and does not require an external network to be accessible, it is not necessary to assign an external network IP to the device at startup, and only the connectivity between the management platform and the device management port can be ensured. Based on the above considerations, a port on the control bridge br-con is ba-DHCP-if, and this port is bound to the DHCP server service to provide local DHCP for the virtual security device. In project realization, DNSmasq is used in this project. DNSmasq is a small and convenient tool for configuring DNS and DHCP, suitable for small networks. The address and related commands assigned by DHCP can be configured to a single host or to a core device (such as a router). DNSmasq supports both static and dynamic DHCP configuration methods. This project uses the dynamic DHCP method, and the interval of the IP address pool to be allocated is 120.0.0.1/24. Ba-router is a Linux network namespace, which is a network namespace. It is used to implement NAT (Network Address Translation) between the virtual security device network and the host network, that is, the IP address segment allocated by the local DHCP service. NAT forwarding enables virtual security devices to connect to external

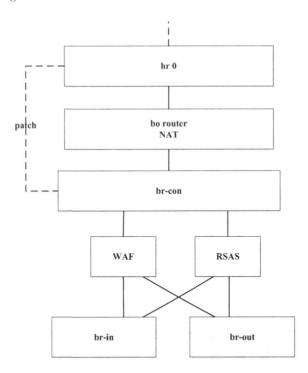

Fig. 10. Management platform network topology

networks. In business, the virtual security device does not have a protection task at the initial startup and does not require external network access [14].

However, when many commercial virtual security devices are launched, they will first go to the cloud of the security vendor to verify the validity of the local license, and then decide whether to provide routine security services. Therefore, NAT in this solution is to ensure that the virtual security device can communicate with the cloud regularly when it is started.

The brO in the topology is a bridge connected to the external network. br-con and brO need to be connected in a patch mode to ensure that after the virtual security device assigns an external network IP, external access traffic can correctly reach the virtual security device.

(3) Database

The information that the device management platform needs to store is mainly device information, and the database selected by this system is Redis. Redis is a remote in-memory database. It not only has stable performance but also has replication characteristics and a unique data model for problem-solving. Redis supports five different types of data structures, and many problems can be stored directly without data conversion.

Besides, through replication, persistence, and client sharding, users can easily extend Redis into a system that can contain hundreds of GB of data and process millions of requests per second.

4   Conclusion

NFV is considered as a foundation of modern networking [19]. It can support various applications such as Internet of things [10,12] and mobile computing [21]. Based on software-defined security architecture and NFV technology, this paper designs and implements a security service enhancement system. The innovation lies in building a bridge between security services, changing the delivery mode of cloud security services, and transforming from the delivery of a single security service to the delivery of multiple security services in a coordinated security protection scheme. The change of the delivery mode not only lowers the threshold for using security services and improves the user experience, but also enhances the security protection efficiency and resource utilization of the cloud platform, and reduces the cost of cloud computing vendors. The orchestration system in this article is deeply integrated with software-defined security technology, maximizing the advantages of software-defined security technology, shielding the differences between various security vendors' devices in the data plane, and weakening the intricate technical details of the underlying system. With which the orchestration engine can focus on the processing of high-level orchestration logic.

Acknowledgement. This research is supported by the Science and Technology Projects of State Grid Jiangsu Electric Power Co., Ltd. (J2019123), NSFC Project No. 61772358 and NSFC Project No. 61572347.

References

1. AT&T Embraces Network Functions Virtualization and May Open Source its NFV Platform. http://alturl.com/mzsyy
2. AT&T FlexWare. http://alturl.com/2c6ev
3. Network Function Virtualization (NFV) Market. http://alturl.com/xps2u
4. Openstack : Security Vulnerabilities. https://www.cvedetails.com/vulnerability-list/vendor_id-11727/Openstack.html
5. Basile, C., Lioy, A., Pitscheider, C., Valenza, F., Vallini, M.: A novel approach for integrating security policy enforcement with dynamic network virtualization. In: Proceedings of the 2015 1st IEEE Conference on Network Softwarization (NetSoft), pp. 1–5. IEEE (2015)
6. Basile, C., Valenza, F., Lioy, A., Lopez, D.R., Perales, A.P.: Adding support for automatic enforcement of security policies in NFV networks. IEEE/ACM Trans. Netw. **27**(2), 707–720 (2019)
7. Basta, A., Kellerer, W., Hoffmann, M., Morper, H.J., Hoffmann, K.: Applying NFV and SDN to LTE mobile core gateways, the functions placement problem. In: Proceedings of the 4th Workshop on All Things Cellular: Operations, Applications, and Challenges, pp. 33–38 (2014)

8. Farris, I., Taleb, T., Khettab, Y., Song, J.: A survey on emerging SDN and NFV security mechanisms for IoT systems. IEEE Commun. Surv. Tutor. **21**(1), 812–837 (2018)

9. Firoozjaei, M.D., Jeong, J.P., Ko, H., Kim, H.: Security challenges with network functions virtualization. Future Gener. Comput. Syst. **67**, 315–324 (2017)

10. Gong, W., et al.: Fast and adaptive continuous scanning in large-scale RFID systems. IEEE/ACM Trans. Netw. **24**(6), 3314–3325 (2016)

11. Hawilo, H., Shami, A., Mirahmadi, M., Asal, R.: NFV: state of the art, challenges, and implementation in next generation mobile networks (vEPC). IEEE Netw. **28**(6), 18–26 (2014)

12. Liu, H., Gong, W., Miao, X., Liu, K., He, W.: Towards adaptive continuous scanning in large-scale RFID systems. In: IEEE INFOCOM 2014-IEEE Conference on Computer Communications, pp. 486–494. IEEE (2014)

13. Marchetto, G., Sisto, R., Valenza, F., Yusupov, J.: A framework for verification-oriented user-friendly network function modeling. IEEE Access **7**, 99349–99359 (2019)

14. Naudts, B., Flores, M., Mijumbi, R., Verbrugge, S., Serrat, J., Colle, D.: A dynamic pricing algorithm for a network of virtual resources. Int. J. Netw. Manag. **27**(2), e1960 (2017)

15. Ordonez-Lucena, J., Ameigeiras, P., Lopez, D., Ramos-Munoz, J.J., Lorca, J., Folgueira, J.: Network slicing for 5G with SDN/NFV: concepts, architectures, and challenges. IEEE Commun. Mag. **55**(5), 80–87 (2017)

16. Palkar, S., et al.: E2: a framework for NFV applications. In: Proceedings of the 25th Symposium on Operating Systems Principles, pp. 121–136 (2015)

17. Pedone, I., Lioy, A., Valenza, F.: Towards an efficient management and orchestration framework for virtual network security functions. Secur. Commun. Netw. **2019** (2019)

18. Reynaud, F., Aguessy, F.X., Bettan, O., Bouet, M., Conan, V.: Attacks against network functions virtualization and software-defined networking: state-of-the-art. In: 2016 IEEE NetSoft Conference and Workshops (NetSoft), pp. 471–476. IEEE (2016)

19. Stallings, W.: Foundations of Modern Networking: SDN, NFV, QoE, IoT, and Cloud. Addison-Wesley Professional, Boston (2015)

20. Yang, W., Fung, C.: A survey on security in network functions virtualization. In: 2016 IEEE NetSoft Conference and Workshops (NetSoft), pp. 15–19. IEEE (2016)

21. Zhao, Y., Li, J., Miao, X., Ding, X.: Urban crowd flow forecasting based on cellular network. In: Proceedings of the ACM Turing Celebration Conference-China, pp. 1–5 (2019)

An Edge-Assisted Video Computing Framework for Industrial IoT

Zeng Zeng[1(✉)], Yuze Jin[2,3], Weiwei Miao[1], Chuanjun Wang[1], Shihao Li[1], Peng Zhou[3], Hongli Zhou[3], and Meiya Dong[2]

[1] Information and Communication Branch, State Grid Jiangsu Electric Power Co., Ltd, Nanjing, China
zengking913@126.com
[2] Tsinghua Wuxi Research Institute of Applied Technologies, Beijing, China
[3] Digital China Information Service Company Ltd, Beijing, China

Abstract. With the rapid development of industrial demands, the Internet of Things triggers enormous interests by industry and academia. By employing IoT technologies, a large number of problems in the industry can be solved by intelligent sensing, wireless communication, and smart software analysis. However, in applying Industrial IoT to improve real-time and immerse user experiences, we found that compared to traditional application scenarios such as tourism, or daily experiences, industrial IoT applications face challenges in scalability, real-time reaction, and immerse user experiences. In this paper, we propose an edge-assisted framework that fits in industrial IoT to solve this fatal problem. We design a multi-pass algorithm that can successfully provide a real sense of immersion without changing the single frame image visual effect in terms of increasing rendering frame rate. From experimental evaluation, it shows that this edge-assisted rendering framework can apply to multiple scenarios in Industrial IoT systems.

Keywords: Industrial IoT · Edge computing · Rendering

1 Introduction

With the rapid development of ubiquitous sensing, wireless communication and data processing technologies, Internet of Things (IoT) has been widely used in various fields, such as environment monitoring, inventory control, and intelligent transportation [14, 15]. Among all these applications, Industrial IoT is projected to be of high potential in the future. For example, in 2015, China pioneered the concept of "Smart Manufacturing 2025", which aims to equip traditional manufacturing with the Internet of Things to drive a new industrial revolution. Furthermore, many international research groups and industrial application alliances have also been established. Within these organizations, global leaders in the manufacturing, telecommunications, networking, semiconductor, and computer industries collaborate to boost Industrial IoT innovations. According to Global Market Research Report [17], the market size of Industrial IoT is expected to reach 753.1 billion USD by the year 2023.

J. Liu et al. (Eds.): MobiCASE 2020, LNICST 341, pp. 53–67, 2020.
https://doi.org/10.1007/978-3-030-64214-3_4

In particular, Industrial IoT applications can monitor the real-time information of physical assets, processes, and systems by sensors. Thereby it can comprehensively monitor the life cycle of industrial production. Meanwhile, Industrial IoT systems employ modern data visualization technologies such as virtual reality (VR) and augmented reality (AR) to construct a "digital twin" [16] of the physical world. In these systems, real-time sensory data are integrated and visually represented on industrial assets. And workers can operate mechanical equipment, vehicles, and pipelines in a remote manner [3, 5].

However, in order to provide users with a real sense of immersion to perceive and manipulate objects in the virtual world, the design and implementation of Industrial IoT systems still face challenges. On the one hand, among the five sensory systems of humans, vision transmits most of the information to the human brain, capturing most of the user's attention, and the contribution of vision to brain stimulation accounts for about 70% [1]. On the other hand, when users perform remote operations in Industrial IoT scenarios, immersive operation methods will cause serious problems such as motion sickness. These problems are not severe in daily experience or under game scenarios. However, the operation of industrial scenarios will bring severe consequences. For example, 1-s operation delay caused by problems such as motion sickness is likely to cause equipment damage [4, 12].

Motion sickness is caused when the signal received by the human visual system and the stimulation felt by the vestibular system is inconsistent, causing the observer's brain to be confused. In the industrial virtual reality scene, the helmet covers the entire field of vision of the user. In order to deceive the user to the maximum, the field of vision felt by the visual system is no longer the size of a screen but approximates the real world. The entire visual field of the human eye, the picture will change in real-time as the user's head and body move. Therefore, two factors can lead to motion sickness. The first factor is that the Industrial IoT system cannot provide real motion output that matches the virtual reality scene. Moreover, the second factor is that the high-resolution rendering in the system brings the system delay and insufficient frame rate, which leads to a mismatch between visual and sensory motion.

In this paper, we propose an edge assisted rendering framework for Industrial IoT. First, we proposed an Edge-based framework, so that the rendering of video information can be completed on edge. Second, we proposed multi-pass and multi-resolution: MultiPass algorithm. Finally, we conducted experiments based on the Unity 3D engine and evaluated it in many scenarios. This framework is proved to perform well in such immerse industrial IoT applications.

2   Related Work

How to improve user experiences of the industrial information system has drawn research interest since the 1980s. Foveal rendering technology was once proposed in military applications [9, 11]. Kocian and Longridge et al. rendered the area of interest of the eye at high resolution and inserted it into the low-resolution area, but the paper did not accurately describe how to operate the rendering pipeline to reduce the resolution of the surrounding area. Geisler and Perry [13] proposed to use multi-resolution graphics pyramids to generate pictures of different resolutions in real-time so that the shape and size of the foveal area changed. The system can produce high-quality images with low aliasing at high frame rates. Clarke [7] then proposed to simplify the resolution of geometric objects far away from the observer and determine the LOD (Levels of Detail, LOD) hierarchy of the object according to the projection area covered by the geometric object. When the area covered by the object is small, a lower resolution model of the object is used. Otherwise, a higher resolution model is used to draw complex scenes quickly. However, the LOD algorithm produces isotropic degraded geometric objects from different perspectives, which is not always satisfactory, especially when viewing large objects at close range. The above method is still in the initial exploration of foveal rendering for video, which has many shortcomings and is not suitable for the industrial IoT systems especially the cloud based systems [20].

Guenter et al. [8] multi-channel rendered three layers of different resolutions around the center of the line of sight. The angle size of the three layers gradually increased, but the sampling rate gradually decreased. The aliasing phenomenon is caused by reducing the resolution of the picture. The author combined multiple anti-aliasing techniques to eliminate them. The system can accelerate the rendering of desktop HD display 3D scenes 5 to 6 times faster, but the advantage of this method is that it is suitable for existing hardware and easy to implement under the software platform. However, this method does not take into account the visual distribution of human eyes in a virtual reality scene, and serious-time aliasing problems have occurred after implementation. Therefore, this solution does not fully use the GazeRender system. Vaidyanathan and Salvi [2] proposed a single-channel. They bounded memory order-independent transparency algorithm, which is mainly used for fast and high-quality rendering of scenes containing geometric primitives with transparency, but the paper does not attempt to eliminate the effects of foveal rendering. Perceptual aliasing problem, this solution is not consistent with the problem to be solved by the system in this paper. Vaidyanathan et al. [10] proposed a method of decoupling the pixel shading rate and the visibility sampling rate for rendering acceleration in 14 years. This method uses the sampling rate of Full HD in the rasterization visibility detection stage and down sampling only in the coloring stage, thereby improving the overall rendering speed while retaining the details of the object. This method significantly reduces the coloring consumption without the introduction of perceptual aliasing and blurring. However, this method currently has no hardware support and is still in the simulation stage, so it is not suitable for the implementation of this system. Patney et al. [6] explored the implementation of this method in virtual reality scenes in 16 years. In summary, none of the existing foveal rendering technologies are fully applicable to the multi-resolution rendering in today's industrial IoT systems [19].

3   System Design

The system architecture and workflow between modules are shown in Fig. 1.

The foveal rendering module is mainly composed of three parts. The first is the discretization of the sampling rate distribution. The target sampling rate of each point on the screen is determined based on the coordinates provided by the eye-tracking and

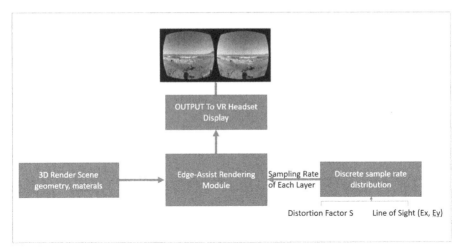

Fig. 1. System architecture

the distribution of human eye acuity. Partitioning and discretization to provide a basis for sampling rates for multi-resolution rendering. After obtaining the sampling rate distribution, the second sub-module of the foveal rendering multi-resolution rendering module performs multi-resolution rendering on the screen according to the sampling rate, and MultiPass implements multi-resolution rendering. The third sub-module is to call the rendering engine to execute a multi-resolution rendering solution and render and output the image to the screen of the virtual reality helmet according to the information of the 3D scene.

This paper mainly introduces the implementation and improvement scheme of the foveal rendering algorithm. The foveal rendering refers to the characteristic that the visual acuity of the human eye is gradually reduced in the visual edge area, and the resolution of the rendered image is gradually decreased from the focus area of the line of sight to the periphery to improve the overall rendering speed. The critical point is to achieve multi-resolution rendering on one screen. This chapter first introduces the implementation of MultiPass (Multi-Pass multi-resolution Rendering). This method can run directly on current GPU hardware without modifying the rendering pipeline. The method is simple and easy to implement. However, after implementing the effect analysis, this paper finds that MultiPass has severe spatial and temporal aliasing in the virtual reality scene due to the down-sampling of the edge layer. Then, for the aliasing problem, we have combined the MultiSampling Anti-Aliasing (MSAA) technology, Temporal Anti-Aliasing technology, and image contrast enhancement directly supported by the hardware into the MultiPass rendering algorithm. The aliasing is reduced to a level almost imperceptible to the human eye. The description of the foveal rendering technology based on the MultiPass algorithm includes the principles and implementation methods of the multi-pass multi-resolution rendering algorithm MultiPass.

3.1 Discretization of Sampling Rate Distribution

According to the idea of sample rate discretization, this paper adopts a three-layer sample rate discretization and nesting and superposition scheme to achieve the concave center rendering, as shown in Fig. 2. Since each layer is rendered with the same resolution and the sensitivity of the human eye is gradually reduced from the center to the periphery, the resolution of each layer must meet the highest resolution requirements of the human eye. People feel a significant change in resolution. The oblique line in Fig. 2 represents the distribution of the minimum resolving angle MAR of the human eye along with the eccentric angle. In this paper, two eccentric angles e_1 and e_2 are selected to divide the screen into three parts: the center layer, the transition layer, and the edge area. The three visual layers use sampling rates are R_0, R_1, R_2, and the corresponding minimum resolution angles are $\omega_0, \omega_1, \omega_2$.

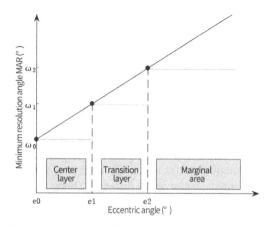

Fig. 2. Parameter selection when the number of eccentric layers is 3.

The selection of volume discretization angle and sampling rate need to consider both human visual effects and rendering performance acceleration. First of all, the visual effect of the human eye needs to be guaranteed. In the sight tracking model in Chap. 3, the concept of sight tracking error δ is proposed. After the experimental measurement, when δ is 1.5, the sight tracking model will ensure that the recognition point is 90% within error δ. In order to compensate for the eye-tracking error, when determining the eccentric angles e1 and e2 of the two layers at the center, it is necessary to increase the angle by δ from the theoretically derived angle. It is to ensure that the area within the e1 and e2 angles around the center of sight of the human eye can always be sampled according to $\omega 0$ and $\omega 1$ Rate rendering. Figure 3 shows a discretization model that compensates for the aforementioned tracking error. Assume that the eccentric angle of e1 is selected as the dividing line, Ex and Ey represent the visual center recognized by the human eye. After the minimum sampling rate ω_0 of the region is determined, due to the existence of human eye recognition error δ, $\omega 0$ will be used as the angular radius as Rendering sampling rate in the range of $\delta + e1$.

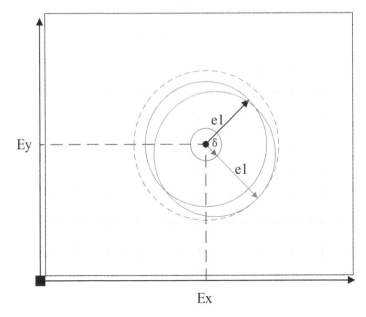

Fig. 3. Sampling rate discretization model to compensate for tracking error.

The relationship between the sampling rate and the minimum resolution angle is described below. Thus, the sampling rate R of each region is calculated according to the minimum resolution angle MAR at which the inner edge is located. In general, the sampling rate can be determined based on the distance D between the human eye and the screen, combined with the minimum resolution angle ω0. However, in a virtual reality helmet, the change in the distance between the screen and the human eye due to eyepiece refraction must also be considered.

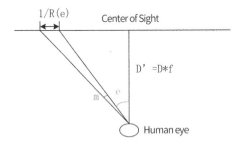

Fig. 4. Relationship between sampling rate and minimum definition view angle.

The sampling rate calculation model is shown in Fig. 4. It is assumed that the refraction of the eyepiece magnifies the screen distance by f times, e is the eccentric angle, ω is the minimum resolution angle MAR at e, and R (e) represents the sampling rate at e. D is the actual distance of the human eye from the screen, and D is the distance between the human eye and the screen after being refracted by the eyepiece. The formula for calculating the sampling rate is derived,

$$\frac{1}{R(e)} = \frac{D'}{(\cos(e))^2} \sin\left(\frac{\omega}{2}\right) * 2 \tag{1}$$

Therefore, the minimum viewing angle MAR is small, so a small angle trigonometric function approximation formula can be used, and the formula 1 can be simplified to,

$$\frac{1}{R(e)} = \frac{D'\omega}{(\cos(e))^2} \tag{2}$$

In summary, if e1 and e2 are selected as the dividing line of the eccentric angle, the sampling rate calculation formula of the three regions is,

$$R_i = \frac{(\cos(e_i))^2}{D'\omega_i} \quad i = 0, 1, 2 \tag{3}$$

According to previous studies [18], in order to ensure the balance between rendering effect and rendering speed, the eccentric angles $e_1 = 5°, e_2 = 30°$ were selected. Because within a 5° viewing angle, the human eye recognizes objects with near full HD visual acuity. The "aliasing phenomenon" occurs in the area from 5° to 30°. That is, the human eye can see some high-frequency information, but cannot clearly see the texture and direction of high-frequency details. From 30° outward, it is the edge area of the human eye. In this visual area, only the low-frequency information of the picture can be seen. According to the research on the distribution of visual acuity of human eyes, the minimum resolution angle of human eyes within 5° is $1/48°$. After the screen, the virtual reality helmet is refracted by the eyepiece. The distance D' from the screen to the human eye is about 25 cm. According to the formula 4–3, the minimum sampling rate of the visual center area is 275PPI. The above is a combination of the classic sampling rate discrete method and the virtual reality environment, which is called the MultiPass algorithm in this paper.

3.2 Implementation of Multi-layer Channel Overlay

This paper mainly introduces the implementation of MultiPass with multi-channel resolution rendering. MultiPass divides the screen into several concentric square layers according to the eccentric angle radius e from the center of the human eye. Each square area is rendered at a gradually lower sampling rate from the inside to the outside, and then multi-layered and nested to form a foveal rendering image.

This paper implements the MultiPass algorithm with three layers of overlay. Around the focus of the line of sight, MultiPass rendered three eccentric layers (the yellow border represents the rendering range of the center layer, the blue border represents the middle layer, and the green border represents the edge layer). The two eccentric layers in the middle cover only a part of the display area, and the edge layer covers the entire display area. The center layer is rendered at the highest sampling rate (consistent with the display resolution). The sampling rate of the two viewing areas outward from the center layer is obtained from the sampling rate calculation formula introduced above.

The three channels of MultiPass render the covered display area separately and then superimpose them into a high-definition image with blurred visual edges. But it is worth noting that the overlapping part of each layer is superimposed instead of hollow nested, which causes the overlapping area to be repeatedly rendered more than once, which brings a certain amount of performance waste. However, existing GPUs and rendering pipelines currently only support rendering for uniform sampling rates. Although the idea of hollow nesting has been proposed by researchers [6, 30], there is only a stage of software simulation and theoretical derivation.

MultiPass achieves a multi-sampling rate by simulating three virtual "screens." The resolution of each virtual screen is set to the minimum sampling rate of the region. In the Unity 3D game rendering engine, multiple cameras (Cameras) are nested with each other to implement the above MultiPass algorithm.

In the Unity3D game rendering engine, the Camera control provides users with a field of view to watch the 3D world. It has a renderer packaged inside it that can render 3D models within the field of view according to angle and position. For example, a virtual reality game usually developed based on Unity3D can display the virtual world from different angles and distances by controlling the position and posture of the Camera, and provide users with a picture.

In order to implement the three-layer MultiPass algorithm, three cameras (c1, c2, c3 from the inside to the outside) were used as a set of renderers in the experiment. Then control camera parameters to achieve different resolution levels of rendering (c1, c2, c3 in order from high-definition to low-definition rendering), the parameter list is shown in the following Table, (Table 1)

Table 1. MultiPass camera perimeter in unity 3D.

Camera no.	Rendering view (Eccentric angle diameter)	Rendering sample rate (ratio to full resolution)
C1	120	0.3
C2	60	0.6
C3	10	1

Then link these Cameras together and control their direction through different controls. The outermost c3 represents the field of view of the helmet and is controlled by the helmet's gyroscope. The center of the two cameras is the same, and the coordinate system is the relative coordinate system of c1. Controlled by the line-of-sight tracking module, that is, the center of the camera coincides with the center of the line of sight. When the human eye rotates, the two cameras in the middle also rotate accordingly so that the visual center area is rendered in c1 HD, the transition area is rendered in c2, and the edge area The MultiPass algorithm rendered by c3 low-resolution. In Untity 3D, the algorithm is described as follows (Table 2):

Table 2. Algorithm description

Input: The position of focus updated by gaze tracking

1: FocusPointPosUpdate(Position){

2:   SetFocusPoint(new Vector2(Position.x, Position.y));

3:}

4: //Setting Camara Parameter according to the position of the focus

  5: SetFocusPoint(newFocusPoint, Camera cam){

  6: Left = newFocusPoint.x-cam.width/2

  7: Right = newFocusPoint.y+cam.width /2

  8: Bottom = newFocusPoint.y- cam.height/2

  9: Up = newFocusPoint.y+ cam.height/2

  10: // Set the position of the four vertices of Camara, the distance between the closest rendering and the furthest rendering

  11: cam.projectionMatrix = PerspectiveOffCenter(left, right, bottom, top,

  12: cam.nearClipPlane, cam.farClipPlane)

13: }

3.3   MultiPass Algorithm Analysis

After implementing the MultiPass algorithm in a layered and nested manner, this section analyzes its performance and verifies the effect, and lays the foundation for the improvement of MultiPass in the following. According to the principle and implementation of the above MultiPass algorithm, it is known that the three layers are superimposed on each other. The central layer area is rendered three times by three resolutions, and the middle layer is rendered two times by two resolutions. Wasted rendering performance. Therefore, we need to choose the proper demarcation angle to minimize this kind of performance waste. The following section analyzes the selection of MultiPass eccentric layer parameters.

The visual acuity model MAR slope is defined as m, the horizontal resolution of the virtual reality helmet D^*, the field of view width W^*, the aspect ratio is α^*, and the viewer-to-screen distance V^*. The eccentric layer is represented as L_i, when $i = 1$ represents the center layer, when $i = n$ represents the outermost layer, and the corresponding eccentric angle is e_i.

The eccentric layer sampling factor S_i ,when $S_i \geq 1$, it means that the pixel size of this layer is a multiple of the pixel size of the display unit. When $S_i = 1$, it means that the sampling rate of the center layer is consistent with the sampling rate of the display. The outermost layer is a self-sampling of the entire screen image and maintains the aspect ratio of the display. e^* indicates the maximum field of view of a single eye, ω^* indicates the minimum resolution angle supported by the virtual reality helmet, and is calculated by the display configuration parameters as the following Equation. ω_0 indicates the minimum resolution angle that can be distinguished by the human eye, usually $\omega_0 \leq \omega^*$.

$$\omega^* = tan^{-1}\left(\frac{2W^*}{V^*D^*}\right) \tag{4}$$

The oblique line segment indicates a MAR line with a slope of m. The horizontal line segment represents the selected eccentric layer parameters. The eccentric angle gradually increases from left to right. The eccentric layers are the center layer, the middle layer, and the edge layer. e^* represents the maximum field of view of a single eye, ω^* represents the minimum resolution angle supported by the virtual reality headset, and ω_0 represents the minimum resolution angle that can be resolved by the human eye. The eccentric layer parameter uses a piecewise constant to approximate the MAR linear model (shown in purple horizontal line segments in above Fig. 2. It is a conservative parameter selection method because the parameters are always below the target MAR line. The increasing eccentric angle e_i determines the corresponding sampling factor S_i, that is,

$$\begin{cases} s_{i+1} = \frac{\omega_i}{\omega^*} = \frac{me_i+\omega_0}{\omega^*}, 1 \le i \le n-1 \\ s_1 = 1 \end{cases} \tag{5}$$

The horizontal diameter D_i of each layer in pixels is calculated from the eccentric angle e_i and the sampling factor S_i, that is,

$$D_i = \begin{cases} 2\frac{D^*}{s_i}tan(e_i)\frac{V^*}{W^*}, 1 \le i < n \\ \frac{D^*}{s_i}, i = n \end{cases} \tag{6}$$

The number of pixels rendered per layer is given by

$$P_i = \begin{cases} (D_i)^2, 1 \le i < n \\ \alpha^*(D_i)^2, i = n \end{cases} \tag{7}$$

Finally, minimizing the total pixels gives

$$min(P) = min\left(\sum_{i=1}^{n} w_i P_i\right) \tag{8}$$

Since the range of e^* is constant and known, we discretize the range of e^* and search for the P value at each eccentricity angle $0 < e_1 < e_2 < \cdots < e_{n-1} < e^*$ to find the minimized result.

From the above analysis, it can be known that the MultiPass algorithm can determine the size and sampling rate of each eccentric layer based on simple mathematical formulas and optimization equations, so as to ensure that rendering effects are minimized due to the overlapping of rendering channels waste.

4   Experimental Evaluation and Verification

We conduct an experimental evaluation of a context-aware virtual reality rendering acceleration system.

4.1 Experimental Design

The multi-pass and multi-resolution rendering algorithm MultiPass is implemented based on existing hardware and rendering pipelines, so it can directly compare the rendering frame rate with full-pixel rendering that does not use foveal rendering, thereby representing the improvement in algorithm speed of foveal rendering. Through the comparison of shaded pixels and rendering effects, the improvement over MultiPass is explained. In order to test the above algorithm, the virtual reality scenes selected in the experiment include three as shown in Fig. 5: (a) a game scene with more complex textures and models, (b) a classroom scene with a simple model, and (c) a scene containing two The museum scene of the layer definition model (LowPoly and HighPoly) is used to test the effect of the MultiPass algorithm.

（a）（Gaming Scene）（b）（ClassRoom Scene）(c)（Museum Scene）

Fig. 5. Three virtual reality scenarios used in the experiment

4.2 Foveal Rendering Module Evaluation

When analyzing the performance of foveal rendering, we first compare the rendering sampling rate and sampling points of foveal and full-pixel rendering and then compare the frame rates of MultiPass and full-pixel rendering. Figure 6 shows the comparison of the sampling rate between foveal rendering and full pixel rendering. In the foveal rendering algorithm, we developed the foveal rendering based on the distribution of human visual acuity (black and green oblique lines). Sampling rate distribution scheme (shown in red). The central sampling rate of this experiment is between visible acuity and separable acuity. Experiments have proved that this compromised sampling rate scheme and anti-aliasing can make users unable to perceive edge down-sampling. As can be seen from the figure, compared with full pixel rendering (blue line), the sampling rate of the foveal rendering in the visual transition area and edge area is much lower than that of the full pixel rendering. In general, the sampling rate shown in the figure is used for the foveal rendering. The sampling point in the visual transition area ($5°$–$30°$) is 20–0% less than the original. The visual edge area can be rendered about 80 less than the original image. -90% of sampling points because the number of sampling points is proportional to the square of the sampling rate.

Due to the reduction in rendering sampling points, the rendering rate will be greatly improved. Even if there is a performance waste phenomenon caused by layer overlay in the MulitPass algorithm, the visual edge area occupies more than 60% of the area of the map, so the rendering speed improvement is still considerable. Figure 6 shows the

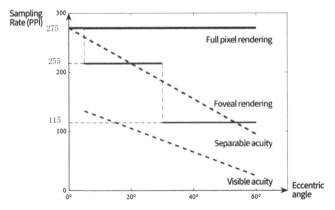

Fig. 6. Comparison of sampling rates between MultiPass and full rendering.

comparison between the full resolution and the foveal rendering algorithm MultiPass in the rendering frame rate in three scenarios. It can be seen that foveal rendering has brought significant performance improvements in various different scenarios. The rendering frame rate has been increased by 2–3 times, all exceeding the preset target value of 120 frames/second.

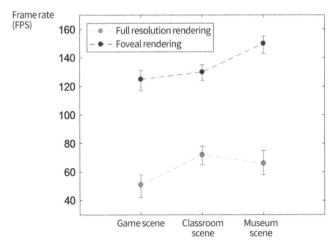

Fig. 7. Comparison of the frame rate of the three virtual reality scenes with MultiPass foveal rendering and full rendering.

In order to help further analyze the performance of the proposed post-processing foveal rendering method in the system, this paper uses wavelet transform to analyze the performance of images at different frequencies and different spatial positions. In this experiment, a two-dimensional Haar wavelet transform is used to analyze the image at different decomposition levels (Image Decomposition Level). The wavelet coefficients of different regions of the image are combined with the eccentricity to analyze the relationship between frequency changes and the angle of sight.

The mother wavelet ψ (t) of the Haar wavelet can be described as

$$\psi(t) = \begin{cases} 1 & 0 \le t < \frac{1}{2} \\ -1 & \frac{1}{2} \le t < 1 \\ 0 & \text{Other} \end{cases} \tag{9}$$

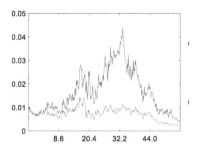

(a)   Eccentricity under HF Wavelet Transform

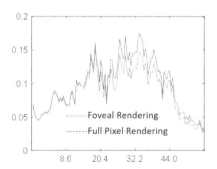

(b)   Eccentricity under IF Wavelet Transform

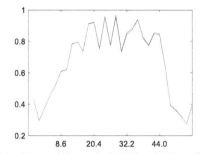

(c)   Eccentricity under LF Wavelet Transform

Fig. 8. Wavelet transform analysis of foveal rendering and full pixel rendering

The zoom function φ (t) can be described as:

$$\phi(t) = \begin{cases} 1 & 0 \leq t < 1 \\ 0 & Other \end{cases} \tag{10}$$

The Haar wavelet coefficients are calculated in the changing eccentricity. As the eccentricity changes, the image information is gradually discarded, and the wavelet coefficient (wavelet energy) also decreases. The larger wavelet coefficients highlight the important feature information that is retained. Figure 7 shows the comparison between foveal rendering and full pixel rendering on high frequency, intermediate frequency and low frequency wavelet transforms. We choose 1×1, 3×3, and 5×5 granularities as high-frequency, intermediate-frequency, and low-frequency wavelets, respectively. The abscissa of the image represents the angle away from the visual center, and the ordinate represents the wavelet energy. The higher the wavelet energy, the more image textures representing this frequency, and the richer the information. In the high-frequency wavelet transform in Fig. 8, as the eccentricity angle increases, the wavelet energy is significantly reduced compared to full-pixel rendering, indicating that high-frequency details are lost in the visual edge region. The foveal rendering of the mid-frequency and low-frequency wavelet energy has less loss than the full-pixel rendering, which indicates that the foveal rendering retains the mid- and low-frequency information of the image at the visual edge.

5   Conclusion

In this paper, we studied a significant framework to improve user experiences in industrial IoT systems. We design a new foveal rendering technology to accelerate the rendering rate to avoid motion sickness in industry.

Given the characteristics of industrial IoT equipment: high refresh rate requirements, large field of vision, and eyepiece distortion, we have conducted a full investigation, demonstration, and improvement of existing related technologies to adapt to the use scenario of virtual reality equipment [18]. We proposed an edge-assist rendering framework to solve the problem. We implemented a multi-pass and multi-resolution rendering algorithm MultiPass, which was based on the Unity3D development engine. In this paper we evaluated its performance and rendering effect. From experimental evaluation, our framework can increase the rendering rate by 2–3 times without affecting the user's visual experience. These results indicate that this research is an effective method to solve the existing problem in terms of user experiences improvement in industrial IoT systems.

Acknowledgement. The paper is supported by the Science and Technology Project of State Grid Corporation of China: "Research on Key Technologies of Edge Intelligent Computing for Smart IoT System" (Grant No. 5210ED209Q3U), NSF China Key Project (Grant No. 61632013), National Key Research and Development Project (Grant No. 2018YFB2200900).

References

1. Lang, D.J.: For virtual reality creators, motion sickness a real issue (2016). http://phys.org/news/2016-03-virtual-reality-creators-motion-sickness.html
2. Salvi, M., Vaidyanathan, K.: Multi-layer alpha blending. ACM SIGGRAPH Symposium on Interactive 3D Graphics and Games, pp. 151–158 (2014)
3. Meiya, D., Jumin, Z., Biaokai, Z., Zhaobin, L.: CLOAK: Visible Touching and Invisible Protecting in Cloud Based IOT System, CBD2018 (2018)
4. Bohil, C.J., Alicea, B., Biocca, F.A.: Virtual reality in neuroscience research and therapy. Nat. Rev. Neurosci. **12**(12), 752–762 (2011)
5. Meiya, D., Jumin, Z., Biaokai, Z., Zhaobin, L.: "CHAMELEON"- hides privacy in cloud IoT system by LSB and CSE. Concurr. Comput.: Pract. Exp. **31**(245)
6. Patney, A., Salvi, M., Kim, J., et al.: Towards foveated rendering for gaze-tracked virtual reality. ACM Trans. Graph. (TOG) **35**(6), 179 (2016)
7. Clark, J.H.: Hierarchical geometric models for visible surface algorithms. Commun. ACM **19**(10), 547–554 (1976)
8. Guenter, B., Finch, M., Drucker, S., et al.: Foveated 3D graphics. ACM Trans. Graph. (TOG) **31**(6), 164 (2012)
9. Kocian D. Visual world subsystem. Super Cockpit Industry Days: Super Cockpit/Virtual Crew Systems. pp. 97–103 (1987)
10. Wang, J.G, Sung, E., Venkateswarlu, R.: Eye gaze estimation from a single image of one eye. In: IEEE International Conference on Computer Vision, p. 136. IEEE Computer Society (2003)
11. Longridge, T.: Design of an eye slaved area of interest system for the simulator complexity testbed. Area of Interest/Field-of-View Research Using ASPT (1989)
12. Tan, K.H., Kriegman, D.J., Ahuja, N.: Appearance-based eye gaze estimation. Applications of Computer Vision, pp. 191–195 (2002)
13. Geisler, W.S., Perry, J.S.: Real-time foveated multi-resolution system for low-bandwidth video communication. In: Proceedings of SPIE - The International Society for Optical Engineering, vol. 3299, pp. 294–305 (1998)
14. Xufei, M., Xin, M., Yuan, H., Xiang-Yang, L., Yunhao, L.: CitySee: Urban CO2 monitoring with sensors. In: 2012 Proceedings IEEE INFOCOM (2012)
15. Qingping, C., Hairong, Y., Chuan, Z., Zhibo, P., Li, D.X.: A reconfigurable smart sensor interface for industrial WSN in IoT environment. IEEE Trans. Indust. Inform. **10**(2), 1417–1425 (2014)
16. He, Y., Guo, J., Zheng, X.: From surveillance to digital twin: challenges and recent advances of signal processing for industrial IoT. IEEE Signal Process. Mag. **35**(5), 120–129 (2018)
17. More, A: Market Share (2019). https://www.marketwatch.com/press-release/industrial-IoT-market-2019—globally-market-size-analysis-share-research-business-growth-and-forecast-to-2023-market-reports-world-2019-05-03
18. Mao, X., Miao, X., He, Y., Li, X.Y., Liu, Y.: CitySee: Urban CO2 monitoring with sensors. In: 2012 Proceedings IEEE INFOCOM, pp. 1611–1619. IEEE (2012)
19. Liu, K., Ma, Q., Gong, W., Miao, X., Liu, Y.: Self-diagnosis for detecting system failures in large-scale wireless sensor networks. IEEE Trans. Wirel. Commun. **13**(10), 5535–5545 (2014)
20. Chen, Z., Zhao, Y., Miao, X., Chen, Y., Wang, Q.: Rapid provisioning of cloud infrastructure leveraging peer-to-peer networks. In: 2009 29th IEEE International Conference on Distributed Computing Systems Workshops, pp. 324–329. IEEE (2009)

OBNAI: An Outlier Detection-Based Framework for Supporting Network-Attack Identification over 5G Environment

Yi Shen[✉], Yangfu Liu, Jiyuan Ren, Zhe Wang, and Zhen Luo

Northeast Branch of State Grid Corporation of China, Shenyang, Liaoning, China
183835533@qq.com

Abstract. With the development of 5G, network attacking becomes more and more easy. Many system vulnerability are utilized to be attacked via 5G technology. It leads that the network attack frequency turn to high, and the network attack strength turns to strong. Among all network attack identification methods, outlier detection is one of the most important one. It aims to find data which is much different from most of the others. In this paper, we propose an outlier detection based framework to support network-attack identification. It first uses a novel algorithm to construct core point set so as support efficiently outlier detection. Next, it uses a novel index named ZB-Tree to manage these core points. Thirdly, we propose a predictive IP-table to handle and predict suspicious IP addresses. In this way, we could identify most suspicious IP addresses based on the position relationships among different base stations. Theoretical analysis and extensive experimental results demonstrate the effectiveness of the proposed algorithms.

Keywords: IP-Table · Path selection · ZB-Tree · 5G

1 Introduction

The Internet of Things (IoT) and the fifth generation of wireless technology(short for 5G) are restructuring the digital world. However, one problem is network attack becomes more and more seriously. Many system vulnerability are utilized to attack Internet via 5G technology. It leads that both the network attack frequency and network attack strength all turn to high.

In order to avoid network attack as much as possible, many algorithms are used for network-attack identification, such as machine learning, cluster, outlier detection, and so on. These algorithms can be used for identifying different kind of attacks. Among all of them, outlier detection is one type of powerful algorithms. The key idea behind them is an object is regarded as an outlier if it conforms to the unexpected behavior. Since the behavior of attacking traffic

J. Liu et al. (Eds.): MobiCASE 2020, LNICST 341, pp. 68–79, 2020.
https://doi.org/10.1007/978-3-030-64214-3_5

is very different from common traffic, but common traffic usually has similar behaviors, we could use distance relationship among traffics to identify attacking traffics.

In this paper, we study the problem of network attack identification based on outlier detection [1–3]. In order to effectively and efficiently support this problem, we should overcome the above challenges. Firstly, since the speed of network traffic is fast, it is difficult to efficiently find network attack in the premise that most attack could be identified. Secondly, under 5G environment, since the distance among different base station is near, attacker could use different IP to attack hosts via changing their position. Thus, it is difficult to check which traffics are attack traffic via their IP address.

In order to overcome the above challenges, we propose a novel framework named OBI. It first uses historical data to construct a set of new points. In this paper, we call these points as core points. Based on these points, when processing newly arrived traffics, we evaluate the "distance relationship" between core points and newly arrived traffics. If we can find few core points whose distance to these newly arrived traffics are near, we regard them as non-attacking traffic. Otherwise, we check whether these traffic is from an IP address which had generated attacking traffics. If so, we regard it as an attacking traffics. Otherwise, we use an ELM-based algorithm for further identifying. Above all, the contribution of this paper are as follows:

- We propose a novel algorithm to construct core point set. It first selects a group of representative points from the whole data set. Next, we construct a group of core points based on the position relationship among these representative points. We find the overall cost is $\mathcal{O}(N)$.
- We propose a novel index named ZB-Tree to manage these core points. The key idea behind it is we first compute the Z-address of each core point. Next, we use a B-Tree to maintain Z-addresses of these core points. The benefit is we could use, as few as, cost to support range query.
- We propose a predictive IP-table to handle and predict suspicious IP addresses. When an attack is generated from a given 5G base station i, we map it into a hash table for one thing. For another, we predict the new attacking IP addresses based on the location of base stations i. In this way, we can identify most suspicious IP addresses based on the position relationships among different base stations.

The rest of this paper is organized as follows: Sect. 2 gives background, Sect. 3 presents the framework. Section 4 evaluates the proposed methods with extensive experiments. Section 5 is the conclusion.

2   Background

In this section, we review the algorithms about network attack based on outlier detection. Thereafter, we introduce a novel machine learning technique called ELM. Table 1 summarizes the mathematical notations used in the paper.

2.1   Related Works

In this section, we review the algorithms about network attack based on outlier detection. Due to the importance of network attack based on outlier detection, many researchers have studied this problem.

Table 1. The Summary of Notations

Notation	Definition
D	the historical data set
$\mathcal{I}$	the IP address Table
$\mathcal{E}$	the ZB-Tree
$\mathcal{M}$	the machine learning based idtendifyer
G	the grid
c	a cell c of G
G	multi-resolution grid
O_c	a set of objects contained in c
$|O_c|$	the number of objects contained in c

Jamshidi Y et al. studied the application of detecting large-scale attacks by the K-nearest neighbor algorithm which based on lattice theory£the experiments show that algorithm has a high detection rate of attack data.

Kuang L et al. proposed a KNN algorithm combined by Strangeness and Isolation indicators in the application of internet intrusion. It uses the two indicators to calculate the outlier score of the sample, and then combine the two interest group degree score to judge whether the sample to be tested is an attack sample. Experiments show that compared to using the KNN algorithm with a single indicator£the combination of the two indicators has a better detection rate and lower false alarm rate.

Zhang J et al. proposed a network anomaly detection algorithm based on random forest algorithm law which uses a random forest algorithm to find outliers. Experiments show that algorithm has higher detection rates and lower false alarm rates than the other three network anomaly detection algorithms.

Bhuyan M H et al. proposed a network anomaly detection algorithm (NADO) based on outlier detection. The algorithm first uses the K-Means algorithm to cluster the training data set, and find the cluster center of each cluster as reference samples. Next, they calculate the distance of each sample to be test the reference sample as the sample outlier degree score value, if this value is greater than the given threshold then output the attack. Experiments show that algorithm has a better detection rate compared with C4.5 and ID3.

Yan Shaohua et al. studied the application of improved LOF algorithm in intrusion detection. Experiments show that when $K = 120$ and $\varepsilon = 1.0$, the algorithm has a higher detection rate on all types of attacks in KDDCUP99.

Guo Chun proposed a network anomaly detection algorithm named NADCP. The algorithm first uses the K-Means algorithm to cluster the training data set, and find the cluster center of each cluster as reference samples. Combined with the idea of over-sampling to calculate the outlier score of the sample, if the outlier of the sample to be tested is greater than a certain threshold, the output is an attack.

Intrusion detection algorithms usually use the idea of nearest neighbors search [4–6], which makes the algorithm should overcome the challenges of neighbor parameters-setting reasonable neighbor parameters can make the algorithm have better detection test results, but if the parameter settings are unreasonable, poor test results may be obtained. In addition, such methods often need manually set the outlier threshold, and the setting of the outlier threshold directly affects the detection rate and false alarm rate of the algorithm. Therefore, how to set a reasonable outlier threshold is also a major challenge for such algorithms.

2.2   Extreme Learning Machine

In this section, we introduce an efficiently machine learning algorithm named (ELM)(short for extreme learning machine). It is developed by Huang et al. ELM is based on a generalized single-hidden-layer feed. Compared with traditional neural networks, its hidden-layer nodes are randomly chosen. It leads that ELM [7,8] has the ability of providing us with good generalization performance at thousands of times faster than traditional popular learning algorithms(e.g.,. SVM, neural networks, and so on). In addition, it has good *universal approximation capability* and *classification capability*.

$$\beta_{k+1} = \beta_k + \mathbf{P}_{k+1}\mathbf{H}_{k+1}^T(\mathbf{T}_{k+1} - \mathbf{H}_{k+1}\beta_k) \tag{1}$$

$$\mathbf{P}_{k+1} = \mathbf{P}_k - \mathbf{P}_k\mathbf{H}_{k+1}^{\mathbf{T}}(I + \mathbf{H}_{k+1}\mathbf{H}_k\mathbf{H}_{k+1}^T)^{-1}\mathbf{H}_{k+1}\mathbf{P}_k \tag{2}$$

Compared with the traditional ELM, Huang et al. also propose the online sequential extreme [7] learning machine (abbreviated as OS-ELM). it can learn data one-by-one or chunk-by-chunk with fixed or varied size. Thus, it is suitable for processing streaming data.

$$\mathbf{H}_{k+1} = \begin{bmatrix} \mathbf{G}(\mathbf{a}_1, b_1, \mathbf{x}_{\sum N_j+1}) \cdots \mathbf{G}(\mathbf{a}_N, b_N, \mathbf{x}_{\sum N_j+1}) \\ \vdots \quad \cdots \quad \vdots \\ \mathbf{G}(\mathbf{a}_1, b_1, \mathbf{x}_{\sum N_j+1}) \cdots \mathbf{G}(\mathbf{a}_1, b_1, \mathbf{x}_{\sum N_j+1}) \end{bmatrix}_{N_{k+1}\times l} \tag{3}$$

Given a set of samples (x_i, t_i), where $x_i = [x_{i1}, x_{i2},...,x_{in}]^T \in \mathcal{R}^n$ and $t_i = [t_{i1}, t_{i2},...,t_{in}]^T \in \mathcal{R}^m$, OS-ELM first selects the activation function, the hidden node number and so on. Then, OS-ELM is employed in a two-phase method including: (i) initialization (ii) sequential learning.

$$\mathbf{T}_{k+1} = \begin{bmatrix} \mathbf{T}_{(\sum_{j=0}^k N_j)+1} \cdots \mathbf{T}_{(\sum_{j=0}^k N_j)} \end{bmatrix}^T_{N_{k+1}\times m} \tag{4}$$

In the initialization phase, OS-ELM uses a small set of samples for training. The second phase employs the learning in a chunk-by-chunk way. To be more specific, in the k-th chunk of new training data, OS-ELM firstly computes the partial hidden layer and the output matrix $\mathbf{H}_{k+1}$. Lastly, OS-ELM computes the output weight matrix β_{k+1}. Here, β_{k+1}, $\mathbf{H}_{k+1}$, and $\mathbf{T}_{k+1}$ are computed according to Eq. 5 to Eq. 4.

3   The Framework ODNAI

In this section, we propose an outlier detection based framework to identify network attack. We first discuss the basic idea. Secondly, we discuss how to construct the cluster. Thirdly, we discuss the network attack based on outlier detection.

3.1   The Basic Idea

Formally, let O be a set of historical network data. We carefully select part of them as *verification objects*. Our goal is using these objects to evaluate which newly arrived objects are outliers. In order to find high quality objects, we propose a *local-density* based algorithm, that is, if a region contains many objects, we select few of them as verification objects. Based on these verification objects, we further select fewer objects as *core-objects*. In Sect. 3.2, we will discuss the *core-object* set construction algorithm.

Algorithm 1: The Framework Overview

   Input: IPBase $\mathcal{I}$, CSet $\mathcal{C}$, object o
   Output: type c
1 bool rst r ←searchCore($\mathcal{C}$, o) ;
2 **if** rst *is not outlier* **then**
3 | return;
4 **else**
5 | bool bIn r ←searchIP($\mathcal{I}$, o) ;
6 | **if** bIn *is false* **then**
7 | | bool bIn r ←ELMID($\mathcal{E}$, o) ;
8 | | **if** rst *is true* **then**
9 | | | $\mathcal{I}$ ← insertion($\mathcal{I}$, o);
10 | | | $\mathcal{I}$ ← prediction($\mathcal{I}$, o);
11 | return;

The function of *core-objects* is when a newly arrived object o traffic into the system, we compute the distance between them and *core-objects*. If we find that the distance between o and *core-objects* are all longer than a threshold, we regard it as an outlier, and further evaluate whether it is a network attack based on the IP-table. We will discuss the *core-objects* maintenance and how to quickly evaluate whether o is a network attack in Sect. 3.3.

The IP-Database maintains a group of suspicious IP address. In the 5G environment, since network attackers could continuously change their locations so as to exchange their IP address, suspicious IP in the IP-Database may be frequently changed. In this paper, we propose a hash based index, which is used to maintain suspicious IP addresses. In addition, we should predict new suspicious IP addresses that may appeared in the future.

If a traffic is identified as an attack traffic, but it is not contained in the IP cache, we use a machine learning algorithm named ELM for further identifying it. Here, ELM is proposed by Huang et al. It is based on a generalized single-hidden-layer feed. Compared with neural networks, its hidden-layer nodes are randomly chosen instead of iteratively tuned. The benefit of ELM is it could provide us with good generalization performance at thousands of times faster than traditional popular learning algorithms. Thus, it is suitable for identifying large scale attacking traffic. Since ELM is widely used in many application, we do not explain the ELM-based identifying algorithm in this paper.

As is shown in Algorithm 1, when an object traffics into the system, we first evaluate whether it is an outlier(See line 1). If not, we regard it as an inlier, and do not process it. Otherwise, we should check whether its corresponding IP address is contained in the suspicious IP addresses table. If the answer is yes, we regard it as an attack traffic. Otherwise, we use a machine learning based method to further process it. If the identification result shows that o is a attack traffic, we insert the corresponding address of o into the suspicious IP addresses table $\mathcal{I}$. In addition, we predict which base station will be used by attackers based on the location information of the current attacker.

3.2 The Outlier Based Identification

In this section, we first discuss how to construct the core set. Next, we discuss searching on core object set.

The Core Set Construction. In this section, we select a set of *non-attack* objects to construct Core Set. Let D be a set of *non-attack* objects with size N. The core set construction contains two steps, which are *mapping* and *selection*. In the first step, we construct a grid G, and map all the objects into G according to their coordinates. Here, the side-length of each $c \in G$ equals to r.

After mapping, we compute the number of objects in each cell. Based on the computational result, we find all cells in G whose scores are larger than k. Here, the score of a cell equals to the number of objects contained in it. For these cells, we randomly select k objects in each cell, and delete the others. Lastly, let O_c be a set of objects contained in the cell c. We compute the center of objects in c, use computing result as core points.

Take an example in Fig. 1. Black points refers to the historical objects. We first partition the whole space into $8 * 8$ cells, and then map these objects into

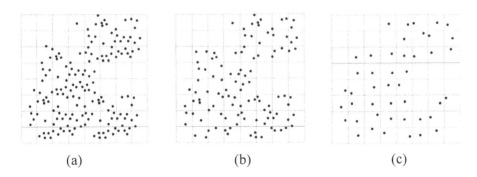

Fig. 1. Problem Definition($k = 3$)

these cells. Let the threshold k be 3. We select all these cells with scores larger than 3, and randomly delete parts of objects contained in them, i.e., the grey cells. The deletion result is shown in Fig. 1(b). Lastly, we compute the core points based on these selected objects. The result is shown in Fig. 1(c).

Indexing Core Set. Once the core set is constructed, we are going to index these core objects. Since core objects are generated by grid, we propose a Z-address based index named ZB-Tree (Fig. 2).

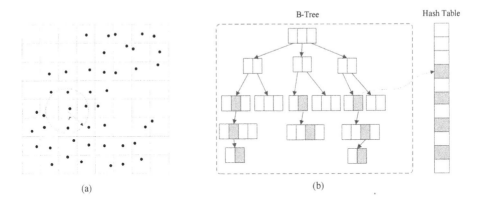

Fig. 2. Indexing Core Points

As is depicted in Fig. 3, ZB-Tree is a Z-curve [2,3] based two-level index. The first level of ZHB-Tree is a B-Tree, which is used to maintain location relationship among all non-empty cells. Since the Z-address of each non-empty cell is an integer, we use a B-Tree to maintain these non-empty cells. The second level of ZHB-Tree is a hash table, which is used to maintain the Z-address of these cells. Since the concept of Z-address has been discussed in many works, we skip the details for saving space.

Searching on ZB-Tree. In this section, we discuss the searching algorithm. Let o_{in} be a newly arrived object. The intuition behind it is if o_{in} is contained in the non-empty cell, it means the distance between it and many objects contained in the historical data set is near. Thus, it should be regarded as a non-attack traffic. In addition, if there exists few non-empty neighbour cells of o_{in}, we should assert that the distance between it and many objects contained in the historical data set is near. Under these cases, we also could assert that o_{in} is not an attack traffic.

Specially, we first compute the Z-address of o. Next, we access the hash table to find whether existing a non-empty cell whose Z-address is maintained by the hash table H. If the answer is yes, the searching algorithm is terminated. Otherwise, we compute the Z-addresses of o's neighbour cells. Based on the computing result, we search on the B-Tree to find whether these neighbour cells are maintained by this B-Tree. If so, we also regard o as a non-attack traffic. Otherwise, we regard it as an attack traffic.

3.3   The IP-Cache Algorithm

As is depicted in Sect. 3.1, if a traffic is identified as an attack traffic, its corresponding IP address is maintained by the hash table H. Since the attack host may be frequency changed, we only maintain the IP addresses of hosts they generate attack traffic in a few hours. In order to achieve this goal, we use a B-Tree to maintain the last attack moment of each attack IP address. In this way, we could know which attack host could be removed from the IP-Cache table based on its last attacking moment.

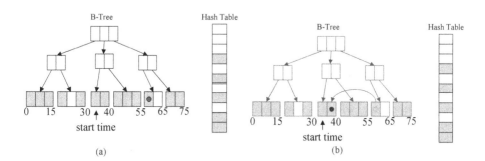

Fig. 3. Indexing Core Points

The IP-Cache Structure. Specially, we use a hash table to maintain the IP address for one thing. For another, we use a B-Tree to maintain the last attack moment of each attack IP address. When a new attack is appeared, we insert the moment of corresponding IP address according to the start time. When an traffic is repeatedly attacking the host, we remove the reference of IP address from its new position. In particularly, if a given IP address has not attack to the host period of time, we remove it from both the hash table and the B-Tree at the same time.

Take an example in Fig. 3. We use the hash table H to maintain attack IP. In addition, we use a B-Tree to maintain the reference of these IP based on their last attacking moment. In addition, the current moment is 35, and we monitor the attacking moment in the last 75 min. The red points refers to an IP which has attack the host 75–30 min before. When a new attacking is appeared, since its corresponding IP address equals to the IP address of the red point, we remove it to another node, i.e., the node with time stamp region [30,40].

The Prediction Algorithm Based on IP Cache. In 5G environment, since the distance among different base-station is short, attacker has the ability of frequently changing IP address. Aiming to this problem, we propose a prediction algorithm that is used for predicting which base-stations attackers may use.

Specially, we first find the location of the base-station b based on the IP address. Next, we submit a range query to find another base-stations whose distance to is nearer than 1km, and obtain these base-stations' IP address. Lastly, we insert these IP addresses into the IP Cache. Since the range query algorithm is simple, we skip the details for saving space.

4   Experimental Evaluation

In this section, we conduct extensive experiments to demonstrate the efficiency of OBNAI. The experiments are based on one real data set and one synthetic dataset respectively. In the following, we first explain the settings of our experiments, and then report our findings.

4.1   Experimental Setting

Data Set. In total, two datasets are used in our experiments, two real data set namely KDDCUP99 and NSL-KDD. The data set is generated from the MIT Lincoln Laboratory, which is used for network detection evaluation. This data set contains roughly 5000000 tuples via simulating the network environment over 7 weeks. Among them, we use 2000000 tuples as training data. In order to apply these two data sets under intrusion detection research, Sal stolfo and Wenke Lee used data mining to analyze its characteristics. Therefore, it is the most widely used data set.

NSL-KDD is oriented from KDDCUP99. Compared with KDDCUP99, it removes redundant tuples from KDDCUP99. In addition, it makes tuple amount difference among different type of tuples small. In this way, it could solve the problem of in-balance. The benefit is it could make the accuracy of training result turns to high.

Metrics. In our experiments, we measure the following metrics by varying different parameters of the system, which are *detection rate*(short for DR), *false* *alarm rate*(short for FAR) and the *accuracy*(short for ACC). Here, they are computed based on the following three equations. Here, P refers to the non-attacked tuple amount, N refers to attacked tuple amount, TP refers to true positive, FN refers to false negative, and FP refers to false positive. In addition, $TP + FN$ equals to P, and $FP + TN$ equals to N.

$$DR = \frac{TN}{N} \tag{5}$$

$$FAR = \frac{FN}{P} \tag{6}$$

$$ACC = \frac{TP + TN}{P + N} \tag{7}$$

Comparisons. we compare the results of OBNAI with a baseline algorithm named BNAI. Compared with OBNAI, it use a baseline outlier detection algorithm. In addition, it does not predict which IP address may generate attack traffics in the future.

4.2 Algorithm Performance

We first evaluate the impact of parameter r to these two algorithms' running time. We find that, with the increasing with r, the running time of these two algorithms are all turning to high. However, compared with the baseline algorithm, the running time of OBNAI increases much lower. The reason behind it is, we construct a small number of core points. In this way, we could use these points for evaluating which ones could be regarded as outliers. Compared with the baseline algorithm, its running time must be lower. Another reason is we use predict algorithm to evaluate which IP address may generate attacked traffics. Thus, we could avoid using machine learning algorithm for further identifying (Figs. 4 and 5).

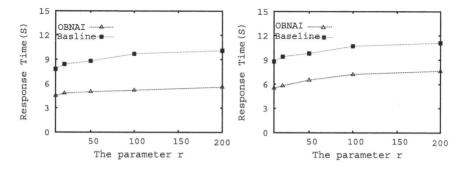

Fig. 4. Running time comparison under different r

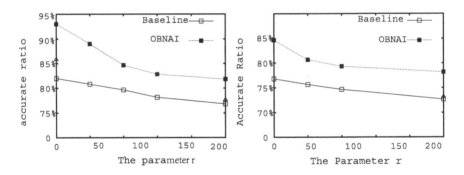

Fig. 5. Accurate ratio comparison under different r

Next, we evaluate the impact of r to the accurate ratio. We find that with the increasing of r, the accurate ratio of these two algorithms all turn to low. The reason behind is when r turns to large, many inliers are wrongly regarded as outliers. However, we also find that when r is large enough, the accurate ratio of OBNAI is not reduced a lot. The reason behind it is, we use a machine learning algorithm for further identifying attacked traffics. Therefore, even a record is wrongly identified a attacked traffic, we could use machine learn algorithm for further verification.

5   Conclusions

In this paper, we propose an outlier detection based framework to support network-attack identification. It first uses a data compression algorithm to reduce data set scale, and then use a Z-order based B-Tree to maintain these core points. The benefit is we could use, as small as, cost for outlier detection. Based on the detecting result, we propose a predictive IP-table based algorithm to find suspicious IP addresses. Last of all, we use a machine learning algorithm named ELM for further verifying suspicious traffics.

References

1. Tran, L., Fan, L., Shahabi, C.: Distance-based outlier detection in data streams. PVLDB **9**(12), 1089–1100 (2016)
2. Juhua, P., Wang, Y., Liu, X., Zhang, X.: STLP-OD: Spatial and temporal label propagation for traffic outlier detection. IEEE Access **7**, 63036–63044 (2019)
3. Zhu, J., Wang, Y., Zhou, D., Gao, F.: Batch process modeling and monitoring with local outlier factor. IEEE Trans. Control Syst. Technol. **27**(4), 1552–1565 (2019)
4. Georgiadis, D., Kontaki, M., Gounaris, A., Papadopoulos, A.N., Tsichlas, K., Manolopoulos, Y..: Continuous outlier detection in data streams: an extensible framework and state-of-the-art algorithms. In: Proceedings of the ACM SIGMOD International Conference on Management of Data, SIGMOD 2013, New York, NY, USA, 22–27 June 2013, pp. 1061–1064 (2013)

5. Hawkins, D.M.: Identification of Outliers. Monographs on Applied Probability and Statistics. Springer, Heidelberg (1980). https://doi.org/10.1007/978-94-015-3994-4
6. Knorr, E.M., Ng, R.T.: Algorithms for mining distance-based outliers in large datasets. In: Proceedings of 24rd International Conference on Very Large Data Bases, VLDB 1998, New York City, New York, USA, 24–27 August 1998, pp. 392–403 (1998)
7. Rong, H.J., Huang, G.B., Sundararajan, N., Saratchandran, P.: Online sequential fuzzy extreme learning machine for function approximation and classification problems. IEEE Trans. Syst. Man Cybern. **39**, 1067–1072 (2009)
8. Huang, G.-B., Zhou, H., Ding, X., Zhang, R.: Extreme learning machine for regression and multiclass classification. IEEE Trans. Syst. Man Cybern. **42**, 513–529 (2012)

Advanced Data Analysis

Image Classification of Brain Tumors Using Improved CNN Framework with Data Augmentation

Xin Ning[1], Zhanbo Li[2], and Haibo Pang[1(✉)]

[1] School of Software Technology, Zhengzhou University, Zhengzhou 450002, China
`innerpeacenx@163.com, panghbzzu@163.com`
[2] Department of Network Management Center, Zhengzhou University, Zhengzhou 450001, China
`iezbli@zzu.edu.cn`

Abstract. At present, the problem of shortage of medical human resources can be solved through mobile medical equipment, the main method to improve the diagnostic performance of mobile medical equipment is to improve the accuracy of the algorithm. Brain tumor classification is to determine the tumor type of patients. The accurate brain tumor classification algorithm can improve the diagnostic performance of mobile medical equipment while assisting doctors in diagnosis. This paper proposes a multi-grade brain classification system using improved CNN framework with extensive data augmentation for differentiating among glioma, meningioma and pituitary tumors, which from three prominent types of brain tumor. First, we locate the tumor and extract the region of interest (ROI). Secondly, to solve the problem of insufficient data samples in the brain tumor classification, we use data augmentation techniques to augment the data samples. Thirdly, VGG-19 and Inception V3 model are improved, and the CNN model is optimized by Adam algorithm. Finally, the improved CNN framework is trained and classified with augmented dataset. Experimental results show that the system proposed in this paper based on data augmentation and improved CNN framework has better classification performance than traditional classifier, and this system can effectively solves the problem of low accuracy caused by insufficient data samples.

Keywords: Mobile medical equipment · Multi-grade brain tumor classification · Data augmentation · CNN · MRI · Adam

1 Introduction

Mobile medical equipment uses mobile communication technology and mobile internet to provide medical services and information. Mobile medical treatment can help improve medical problems. Accurately diagnosing patients through mobile medicine is one of the main problems at present. Brain tumor is one of the most dangerous cancers, the

J. Liu et al. (Eds.): MobiCASE 2020, LNICST 341, pp. 83–101, 2020.
https://doi.org/10.1007/978-3-030-64214-3_6

World Health Organization (WHO) divides brain tumors into four grades I (benign) to IV (malignant). Accurately diagnosing the type of brain tumor can help doctors treat patients. The characteristics of malignant tumors are usually determined by histopathology, but different pathologists have different views on these characteristics, so magnetic resonance imaging (MRI) is introduced. MRI provides a higher contrast for imaging of brain soft tissue.

Deep learning is widely used in brain tumor segmentation, classification, and localization of lesions. Laukamp et al. [1] established a deep learning model that can perform full-automatic detection and segmentation of meningioma against two different MRI images (FLAIR and T1CE). Kamnitsas et al. [2] proposed a dual pathway, 11-layers deep, three-dimensional Convolutional Neural Network for the challenging task of brain lesion segmentation. The author devised an efficient and effective dense training scheme which joins the processing of adjacent image patches into one pass through the network. This method improved the computational efficiency of 3D medical scan processing. Kharrat et al. [3] proposed a hybrid scheme that uses genetic algorithms (GA) and support vector machines (SVM) to classify brain tumor into normal, benign and malignant. In order to solve the problem of insufficient and unbalanced clinical data of meningioma images, Zhu et al. [4] used the LeNet-5 network and oversampling technology to classify meningioma into three categories. Afshar et al. [5] proposed a CapsNets for brain tumor in addition. Zia et al. [6] proposed a generalized classification system for brain tumor based on rectangular window image clipping. This system utilized discrete wavelet transform and PCA, and used SVM as the classifier; the system had low computational complexity, and it was conducive to the development of a universal CAD system. It can be used in any clinical or institution to help radiologists to understand brain tumor MRI images.

Transfer learning is a common method for training large-scale models. The small amount of data can easily cause the model overfitting, and transfer learning is often used to solve such problems. Swati et al. [7] proposed a fine-tuning strategy based on transfer learning using a pre-trained deep convolutional neural network, which performed well on the CE-MRI dataset with an average accuracy of 94.82%. Deepak et al. [8] used pre-trained GoogLeNet to extract features from MRI images by using transfer learning to classify three brain tumors. The research in this paper showed that transfer learning was effective in solving the shortage of data.

Data augmentation is mainly divided into image processing techniques and generative adversarial networks (GAN) [9]. Chang et al. [10] used residual CNN to determine the status of Isocitrate Dehydrogenase (IDH) in low and high gliomas. To increase the size of the training set and prevent overfitting, the author augmented the training set images by introducing random rotations, translations, flips, shearing, and zooming. Han et al. [11] proposed a progressive growing of generative adversarial networks (PGGAN) to solve the problem of insufficient data in real image distribution.

Glioma, meningioma and pituitary tumor have the highest incidence among all brain tumors. For these three types of brain tumors with the highest incidence, how to classify them accurately is a hot topic of current research. The main problems now are as follows: 1) Large-scale labeled medical image datasets are difficult to obtain. 2) Training a large-scale deep learning model with a small amount of data easily leads to model overfitting.

3) Directly using classic deep learning networks for medical image processing may lead to low accuracy of the model, so it is necessary to improve the model.

In this paper, we propose a new framework to classify three pathologic types of brain tumors (glioma, meningioma and pituitary tumor). First, we segment the tumor ROI from the MRI image. Then we use data augmentation technique to augment the data samples. In this paper, we compare two data augmentation methods, one based on image processing technique and the other based on deep convolutional generative adversarial networks (DCGAN) [12]. Finally, we improve the performance of the model by adjusting the network framework and parameters of the model, and use the Adam [13] algorithm to optimize the improved model. The major contributions of this paper are listed below:

1. Due to the lack of public brain tumor datasets, we use two data augmentation methods to augment the data, One is using different image processing techniques to add noise and transformation to the data, and another is using DCGAN to generate new data, and we analyze the impact of the two methods on the model performance.
2. Two improved deep CNN models are proposed. The first model is the improved VGG-19 model, and we improve its network structure. The second model is the improved the Inception V3 model, we use inductive transfer learning to improve it. And using Adam algorithm to optimize the two models. The improved two models have a stronger ability to learn the features of brain tumor data.
3. The effects of different learning rate and batch size on model accuracy and convergence speed are studied. The influence of parameters in the Adam algorithm on the model convergence speed is studied.
4. The two improved models have high specificity, both exceeding 90%, which can effectively judge the authenticity of negative patients.
5. Our diagnostic system has a higher classification accuracy than other methods, so the system is suitable for mobile medical equipment, it can improve the diagnostic accuracy of doctors, and can be used for the positive diagnosis.

2 Method

The design framework of this paper mainly includes three parts: 1) ROI segmentation. 2) Data augmentation. 3) Training model and classification. The system framework is shown in Fig. 1.

The first step is to locate the tumor area of the sample data in the original dataset, segment the tumor, and then obtain the ROI image of the tumor. The second step is to augment the dataset. Expanding a image of the original dataset to 32 images. The size of the datasets after the two data augmentation methods are the same. The third step is to improve and optimize two classic deep learning CNN models, the VGG-19 model and the Inception V3 model. Then using the augmented datasets to train two improved models and classify brain tumor.

2.1 Brain Tumor ROI Segmentation

Accurately locating ROI or lesions in MRI images is a key part of a doctor's diagnosis. The experiments in this paper use the publicly available figshare dataset [14], this dataset

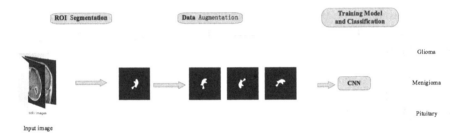

Fig. 1. System framework

contains the tumor boundary coordinates of each data sample, as well as a binary image labeled with the tumor area. We use the threshold segmentation technique proposed by Otsu et al. [15] to segment the ROI of 3 types of brain tumors, and the binary images are generated from the original MRI grayscale images, and white represents the tumor area, as shown in Fig. 2.

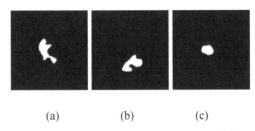

Fig. 2. Images of three brain tumors after ROI segmentation (a: Gliomay. b: Meningioma. c: Pituitary tumor.)

2.2 Data Augmentation

The quality and quantity of training samples are important factors that affect the classification performance of deep learning models. In this paper, we use image processing techniques and DCGAN to augment the dataset.

Data Augmentation Based on Image Processing Technique

Xue et al. [16] used fixed scaling, rotation plus flip and shearing to augment the images. Training deep CNN requires a large-scale dataset, so we use a variety of image processing techniques to achieve data augmentation. A total of 9 image processing techniques, including 5 methods of data augmentation through geometric change techniques: flipping, rotation, shears, skewness and distortions. Considering that when the number of samples increases to a certain extent, it will affect the performance of the model and reduce the robustness of the model. We need to add noisy data samples to the dataset to verify the robustness of the model. So we add noise to the data samples through the four noise addition techniques of gaussian blur, sharpening, edges detection and median blur. Table 1 shows the above 9 data augmentation techniques and 32 parameters, the

Table 1. Different data augmentation techniques with their respective parameters

S. no.	Data augmentation technique	Parameters	
1	Rotation (angle)	90°	
		45°	
		−45°	
		−90°	
		180°	
2	Flip	Level	
		Vertical	
3	Shear	Left 15°	
		Right 15°	
4	Skew	Bottom	0.5
			1.0
		Right	0.5
			1.0
5	Gaussian blur (σ)	0.5	
		1.0	
		1.5	
		2.0	
6	Distortions (grid)	Width = 5.0 height = 10.0	
		Width = 10.0 height = 5.0	
7	Sharpen (lightness)	1.0	
		2.0	
8	Median blur (kernel)	5×5	
		7×7	
		11×11	
9	Edge detection	Direction = 0.5	$\alpha = 0.25$
			$\alpha = 0.50$
			$\alpha = 0.75$
			$\alpha = 1.0$
		Direction = 1.0	$\alpha = 0.25$
			$\alpha = 0.50$
			$\alpha = 0.75$
			$\alpha = 1.0$
			$\alpha = 0.25$

second column refers to augmentation techniques and the third column is the parameters of each technique. Here are a total of 32 parameters, that is, one sample in the original dataset can be expanded to 32 samples.

Data Augmentation Based on DCGAN

GAN is used to generate new data samples, while DCGAN applies the concept of CNN to GAN. In DCGAN, the discriminant network is a CNN. After several layers of convolution, the input image gets a feature, which is then inputted into the logistic function. The output of the function can be regarded as a probability. Compared with the GAN, DCGAN has the following advantages: 1) The network framework of DCGAN is relatively stable. 2) DCGAN can generate high-quality images. We use DCGAN to augment the dataset. During training, when it is found that some abnormal images are generated, these images will be deleted. The data samples generated after training are shown in Fig. 3.

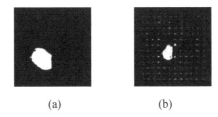

(a)   (b)

Fig. 3. The images generated by DCGAN (a: Normal image. b: Anormal image.)

2.3   Network Model Optimization

In this paper, we improve two deep learning networks and extract features and classification. First, the VGG-19 [17] network and the Inception V3 [18] network are improved respectively, and then adam is used to optimize the two models.

VGG-19 Network Improvement

VGG-19 network consists of 19 weighted layers, in which there are 16 convolutional layers (CONV) and 3 fully connected layers (FC), as shown in Fig. 4 (a). The improved network framework in this paper is 17 weighted layers, with 15 convolutional layers and 2 fully connected layers, as shown in Fig. 4 (b).

The specific improvement steps are as follows: 1) In the original VGG-19 network, a convolutional layer with 64 channels is added after the second convolutional layer, and a convolutional layer with 128 channels is added after the fourth convolutional layer. The purpose is that because the segmented ROI image is a binary grayscale image, the contour features of the tumor are very important, so adding a convolutional layer near the input layer can better extract low-order features. 2) Deleting the 11th, 12th, 15th, and 16th convolutional layers, which can speed up model convergence. 3) Deleting the last full connected layer and using a convolutional layer with output 3 as the new full connected layer, the size of the convolutional kernel is 1×1. And in order to prevent and reduce overfitting, dropout [19] optimization scheme is added after 2 fully connected layers and the last 1 convolutional layer. BP neural network algorithm is used to update and learn the parameters in the CNN framework. After the completion of one epoch, some neurons will be deleted randomly during the next epoch until the training is completed. Figure 4 shows the VGG-19 network framework before and after improvement.

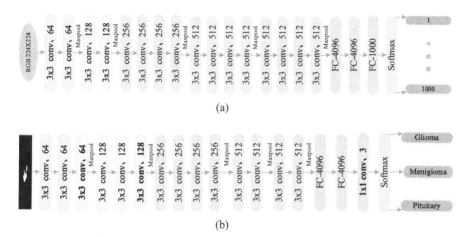

Fig. 4. VGG-19 network framework before and after improvement (a: Original VGG-19 network framework. b: Improved VGG-19 network framework)

Inception V3 Network Improvement

The improvement of Inception V3 in this paper uses inductive transfer learning. In inductive transfer learning, the learning tasks of the source domain and the target domain are different, and the feature space and the edge probability distribution of the source domain and target domain are the same. Figure 5 shows the principle of inductive transfer learning in this paper. We use the imagenet dataset to train the original Inception V3 model, and transfer the learned knowledge to the classification model in this paper. The source domain is the imagenet dataset, the target domain is the figshare dataset, the source task is 1000-class classification, and the target task is the 3-class brain tumor classification. The knowledge learned from the source domain is the feature of the natural images in the imagenet dataset, such as the color feature and shape feature in its sample.

The improvements to the Inception V3 model in this paper are as follows: 1) Deleting the original fully connected layer of the Inception V3 network and reinsert a new fully connected layer with an output size of three to adapt it to the target domain. Deleting the loss function in the original classification layer and use the new loss function. 2) Using the method of transfer learning, and low-level features of the data in the imagenet dataset are learned by the pre-trained layers from the original Inception V3 model, then using the segmented and preprocessed data samples of the figshare brain tumor dataset to train the improved Inception V3 model for the experiment. We set the learning factors for weights and bias at the new fully connected layer to 10, the purpose is to make the

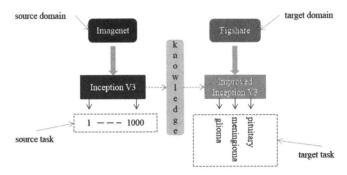

Fig. 5. Model for transfer learning

model learn specific high-level features of the target domain. The convolutional layer of the model is unchanged, only the structure after the FC layer is changed. The Inception V3 network framework before and after the improvement is shown in Fig. 6.

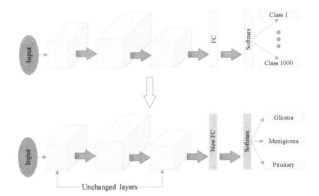

Fig. 6. Improved the inception V3 network framework

Figure 7 shows the improved Inception V3 model classification framework. When training, the model learns classification based on the learned training features and training labels. During testing, the features of the test features are input into the classifier, and the model predicts the test labels. Then calculating the classification accuracy of the test by the actual test label.

Optimizing CNN Networks with Adam

In this paper, we use Adam to optimize two improved models. Traditional CNN networks use stochastic gradient descent (SGD) in the BP framework to minimize the loss function. SGD randomly selects samples to update the gradient at each epoch. During the training process, it will maintain a single learning rate, which will cause each epoch to be not the optimal direction and will cause the loss function to oscillate severely, which will cause the accuracy of the model to decrease. We train the improved VGG-19 model on the original dataset with the SGD optimizer and the Adam optimizer, respectively. The

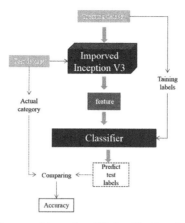

Fig. 7. Improved inception V3 classification framework

average loss function values of the first ten epochs are shown in Fig. 8. In the ten epochs, the changing trend of the Loss value of the SGD optimizer is not obvious, and the Loss value of the Adam optimizer drops rapidly at the third epoch and continues to decline thereafter.

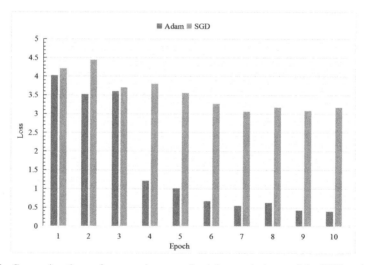

Fig. 8. Comparing the performance between the Adam optimizer and the SGD optimizer

According to Fig. 8, it can be seen that the SGD decline rate is slower than Adam, so we use Adam instead of SGD to minimize the loss function. Adam calculates the adaptive learning rate of each parameter. The process of network parameter update is as follows:

$$g_t = \frac{1}{n} \nabla_\theta \sum_i L(F(Y_i, \theta), X_i) \tag{1}$$

$$m_t = u \times m_{t-1} + (1 - u) \times g_t \tag{2}$$

$$n_t = v \times n_{t-1} + (1 - v) \times g_t^2 \tag{3}$$

First, the Adam initializes the parameter vector, the first moment vector and the second moment vector, and then it will iteratively update each part to make the parameter θ converge. In Eqs. (1), (2) and (3), $L(\theta)$ is the mean square error function, g_t is the gradient function of $L(\theta)$ against θ. m_t is the first moment estimate and n_t is the second raw moment estimate. u and v are exponential decay rates for the moment estimates. At time $t + 1$, updating biased first moment estimate and biased second raw moment estimate.

$$\hat{m} = \frac{m_t}{1 - u_t} \tag{4}$$

$$\hat{n} = \frac{n_t}{1 - v_t} \tag{5}$$

Next, computing bias-corrected first moment estimate and bias-corrected second raw moment estimate. In Eqs. (4) and (5), $\hat{m}_t$ and $\hat{n}_t$ are deviation corrections of m_t and n_t, respectively.

$$\Delta\theta_t = -\eta \times \frac{\hat{m}_t}{\sqrt{\hat{n}_t} + \varepsilon} \tag{6}$$

$$\theta_{t+1} = \theta_t + \Delta\theta_t \tag{7}$$

Finally, the gradient of the objective function for the parameter θ is updated at time $t + 1$, then updating the parameter θ of the model with the computed value above. In Eqs. (6) and (7), η is the step size, ε is the numerically stable small constant. $\Delta\theta_t$ is the updated value of θ_t at time t. At time $t + 1$, adding $\Delta\theta_t$ and θ_t to get the value of θ for θ_{t+1}.

3   Experiments

All experiments are evaluated using NVIDIA GeForceRTX-2080Ti with 64 GB onboard memory and deep learning framework Tensorflow that are installed on Ubuntu 18.04.

3.1   Dataset and Pre-processing

The figshare dataset contains 3064 brain MRI images of 233 patients, and the images belong to the T1-CE MRI. It contains 1,426 MRI images of glioma (89 patients), meningioma contains 708 images (82 patients), and pituitary tumor contains 930 images (62 patients). The resolution of the image is 512×512. The gap between each slice is 1 mm, and the slice thickness is 6 mm. The data is augmented by the data augmentation techniques in Table 1 and DCGAN. In order to compare the impact of the two data augmentation techniques on model performance, the size of the datasets after the expansion of the two data augmentation techniques is the same. After data augmentation, the total number of samples in this dataset is increased from 3064 to 98048, of which glioma images is increased from 1426 to 45632, meningioma images is increased from 708 to 22656, and pituitary tumor images is increased from 930 to 29760.

3.2 Optimizer Setting

For the Adam optimizer, η is step size, the value of the η parameter affects the size of the update value $\Delta\theta_t$ of θ_t at time t, that is, the parameter η affects the descent direction of the loss function and the accuracy. When the value of η is larger, the model cannot converge, and the loss function will always oscillate. When it is small, the model needs to be iterated many times to reach the global optimum. According to the paper [13], we set the exponential decay rate of the moment estimation $u = 0.9$, $v = 0.999$, and the numerically stable small constant $\varepsilon = 10^{-8}$. In order to analyze the effect of η value on model performance, we set different η values, and then train improved network models on the original dataset. Finally, we compare the experimental results.

3.3 Classifier Setting

The VGG-19 model and the Inception V3 model were originally designed for RGB color images, with an input layer of size $224 \times 224 \times 3$. Because the images of the figshare dataset are grayscale images. First, we adjusted the image size, and then copied the grayscale values three times to create three channels. We used patient-level five-fold cross-validation on the dataset before and after the data augmentation to verify the prediction ability of the model. We used the cross-loss function as the loss function of the model.

Different parameter values will seriously affect the performance of the model. Too large learning rate will prevent the loss function from converging. Too small batch size will also make the model unable to converge and the training time will increase, too much batch size will adversely affect the quality of the model, and the computer hardware configuration cannot meet the calculation requirements.

The experimental process is as follows: First, we set different learning rate values and different batch size values. After the dataset was augmented by DCGAN, we trained two improved models separately, compared the accuracy of the models, and analyzed the impact of learning rate and batch size on model accuracy and loss. Secondly, we used the datasets before and after data augmentation to train the two improved models, then we compared the accuracy of the two improved models before and after data augmentation and analyzed the impact of data augmentation on brain tumor classification. Finally, we compared the performance of the two classifiers before and after data augmentation.

3.4 Evaluation Index

This paper conducted the experiments five times, and each experiment followed a patient-level five-fold cross-validation. The loss of the loss function is the average after five experiments. Based on the correct and incorrect classification of the three types of brain tumor, the precision, recall, accuracy and specificity of each brain tumor were calculated according to Eq. (8), where TP, FP, TN and FN are the number of classified cases of true positives, false positives, true negatives and false negatives, respectively.

$$precision = \frac{TP}{TP + FP}$$
$$recall = \frac{TP}{TP + FN}$$
$$specificity = \frac{TN}{TN + FP}$$
$$accuracy = \frac{TN + TP}{TP + TN + FP + FN} \quad (8)$$

4 Analysis of Experimental Results

4.1 Effect of η Parameter on Adam Optimizer

When the two models use Adam optimizer with the original dataset, we set different η values. After 10 epochs, comparing the accuracy and loss of the two models, as shown in Table 2. The experiment result shows: 1) For the improved VGG-19 model, the accuracy is higher when η is 10^{-4}, the loss of the loss function is the smallest, and the convergence speed is the fastest. 2) When the improved Inception V3 model takes different values, the accuracies are not high after 10 epoch. Because the Inception V3 model has more network layers and fewer data samples in the original dataset, so fewer features are learned during training. When the number of epochs is small, the accuracy is low. 3) In the improved Inception V3 model, the accuracy and loss are close when η is 10^{-4} and η is 10^{-5}. When η is small, the model needs a longer training time, so we set η as 10^{-4} for comprehensive consideration.

Table 2. The accuracy and loss of two improved models with different η after 10 epochs

Improved VGG-19 model			Improved Inception V3 model		
η	Loss	Accuracy	η	Loss	Accuracy
10^{-2}	3.604	0.5964	10^{-2}	7.351	0.2666
10^{-3}	1.174	0.6812	10^{-3}	3.638	0.2667
10^{-4}	0.392	0.7911	10^{-4}	1.110	0.5667
10^{-5}	0.403	0.7829	10^{-5}	1.021	0.5625

4.2 The Impact of Learning Rate and Batch Size on Model Accuracy and Loss

Training the model on the dataset augmented by DCGAN, setting different parameters and different epochs, and compare the accuracy and loss of the two models. As shown in

Table 3, according to the results in the table: 1) when the learning rate of the two models are 1×10^{-4} and the batch size of the two models are 64, the accuracy of the two models are the highest and the loss of the loss functions are the smallest. 2) For the VGG-19 model, when the batch size is 64, the learning rate is 1×10^{-4}, and the accuracy of the model differs greatly, indicating that the VGG-19 model is more sensitive to the learning rate. 3) When the learning rate of the two models is 1×10^{-4} and the batch size is 48, 64 and 128 respectively, the difference in model accuracy is small, and the difference in loss value is large, indicating that batch size affects the direction of gradient descent. 4) Comparing when the learning rate is 1×10^{-5} and 1×10^{-4}, it is found that the smaller learning rate will significantly affect the accuracy of the two models.

Table 3. Comparing the accuracy and loss of two improved models with different parameters(where LR, BZ, and E refer to learning rate, batch size and epoch, respectively)

Improved VGG-19 model					Improved inception V3 model				
LR	BZ	E	Acc (%)	Loss	LR	BZ	E	Acc (%)	Loss
1×10^{-3}	32	200	88.69	1.071	1×10^{-3}	32	200	86.43	0.821
1×10^{-4}	48	80	90.68	0.107	1×10^{-4}	48	80	88.09	0.255
1×10^{-4}	64	50	91.73	0.061	1×10^{-4}	64	50	88.96	0.103
1×10^{-4}	128	50	91.12	0.079	1×10^{-4}	128	50	88.77	0.109
3×10^{-4}	64	50	90.35	0.109	3×10^{-4}	64	50	87.33	0.169
1×10^{-5}	64	50	90.30	0.072	1×10^{-5}	64	50	88.82	0.103

4.3   Accuracy Analysis of the Models Before and After Data Augmentation

In the histograms of Figs. 9 and 10, we can see the accuracy of the two models in the original dataset, the dataset augmented with DCGAN and the dataset augmented with 9 augmentation techniques. In the figures, the abscissa is the number of epochs, a total of 50 epochs, taking 10 epochs out of 50 epochs for comparison, and the ordinate is the accuracy. After 50 epochs, the accuracy of the improved VGG-19 on the original dataset is 0.8710, the accuracy of using DCGAN to augment the dataset is 0.9173, and the accuracy of using 9 data augmentation techniques is 0.9109. And the accuracy of the improved Inception V3 model on the original dataset is 0.8531, the accuracy of using DCGAN to augment the dataset is 0.8896, and the accuracy of using 9 data augmentation techniques is 0.8781. At the 20th epoch, the accuracies of the improved VGG-19 model trained with three datasets all reach above 80%, while the Inception V3 model has only the accuracy of dataset augmented by 9 data augmentation techniques reaches above 80% at the 20th epoch. According to the results, data augmentation is effective for training models, and the accuracy of two models using DCGAN is higher than the accuracy of two models using 9 augmentation techniques.

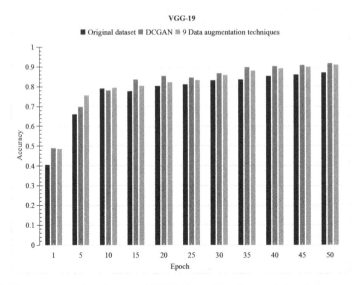

Fig. 9. The accuracy of the improved VGG-19 model before and after data augmentation

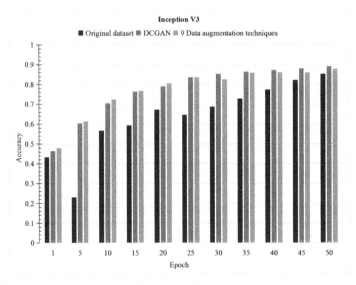

Fig. 10. The accuracy of the improved Inception V3 model before and after data augmentation

4.4 Performance Analysis of Improved VGG-19 Model

Table 4 shows the confusion matrix of the improved VGG-19 model in the original dataset and the dataset augmented using DCGAN, as well as the accuracy. Table 5 shows the classification performance of the improved VGG-19 model before and after data augmentation, including the precision, recall, and specificity of three brain tumors. In Table 4, the accuracy of the improved VGG-19 model before data augmentation is

87.10%. After data augmentation, the accuracy is 91.73%. According to the results in Tables 4 and 5: 1) Before and after data augmentation, the accuracy of the model is increased by about 4.63%, this result indicates that data augmentation has a greater impact on the performance of the improved VGG-19 model. 2) The specificities of the improved VGG-19 model before and after data augmentation are relatively high, both of which are more than 90%, which can effectively judge true negative patients and can be used for the positive diagnosis. 3) Before and after the data augmentation, the precisions of the model for glioma are higher, while the precisions of meningioma and pituitary tumor are lower, which may be caused by the imbalance of the data of the three brain tumors. 4) The precision and recall of meningioma are low, and the proportion of meningioma misclassified is higher than other brain tumors.

Table 4. Confusion matrix of improved VGG-19 model before and after data augmentation (where M, G, and P refer to meningioma, glioma and pituitary, respectively)

Confusion matrix of improved VGG-19 before data augmentation				Confusion matrix of improved VGG-19 after DCGAN				
		Predicted					Predicted		
		G	M	P			G	M	P
Actual	G	1273	81	72	Actual	G	43243	952	1437
	M	60	599	49		M	1369	19484	1803
	P	51	82	797		P	507	2040	27213
Accuracy		**87.10%**			Accuracy		**91.73%**		

In the experiment, we compared the accuracies of the model before and after the improvement on the dataset augmented using DCGAN. The accuracy of the VGG-19 model before improvement is 89.81%, and it is increased to 91.73% after improvement. According to the comparison results, the accuracy of the improved model is increased, indicating that the improvement of the VGG-19 model is effective on this dataset. After improvement, the ability to learn features of the model becomes stronger.

Table 5. Evaluation of improved VGG-19 model before and after data augmentation

Evaluation of improved VGG-19 before data augmentation				Evaluation of improved VGG-19 after DCGAN			
Type	Precision	Recall	Specificity	Type	Precision	Recall	Specificity
G	91.98%	89.27%	93.22%	G	95.84%	94.76%	96.42%
M	78.60%	84.60%	93.08%	M	86.69%	86.00%	96.03%
P	86.82%	85.70%	94.33%	P	89.36%	91.44%	95.26%

4.5 Performance Analysis of Improved Inception V3 Model

Table 6 shows the confusion matrix of the improved Inception V3 model in the original dataset and the dataset augmented using DCGAN, as well as the accuracy. Before and after data augmentation, the accuracy of the improved Inception V3 model increased from 85.31% to 88.96%. Table 7 shows the classification performance of the improved Inception V3 model before and after data augmentation. According to the results in Tables 6 and 7 :1) After the data augmentation, the performance of the improved Inception V3 model has been significantly improved, indicating that using data augmentation can improve the performance of the classifier, but the precision and specificity of glioma are reduced after data augmentation. 2) The accuracy of the improved Inception V3 model before and after data augmentation is increased by 3.65%, compared with the improved VGG-19 model, the increased of accuracy is relatively small. 3) The accuracy and specificity of meningioma and pituitary tumor were less than 90% before and after data augmentation. Comparing the results of the VGG-19 model, it can be seen that the data imbalance affects the CNN model to extract features, and the quality and quantity of the dataset seriously affect the classification performance of the model.

Table 6. Confusion matrix of improved Inception V3 model before and after data augmentation (where M, G, and P refer to meningioma, glioma and pituitary, respectively)

Confusion matrix of improved inception V3 before data augmentation				Confusion matrix of improved inception V3 after DCGAN				
		Predicted					Predicted		
		G	M	P			G	M	P
Actual	G	1229	101	96	Actual	G	41613	2335	1684
	M	55	594	59		M	1884	19353	1419
	P	20	119	791		P	842	2663	26255
Accuracy		**85.31%**			Accuracy		**88.96%**		

The accuracy of the Inception V3 model before improvement is 86.79%, and it is increased to 88.96% after improvement. According to the comparison results, after the improvement, the accuracy of the model is increased, this can indirectly explain that transfer learning is conducive to training a deeper network model.

4.6 Comparison with Other Methods

This section compares our method with other methods which use the figshare dataset to classify the brain tumor. As shown in Table 8, the last two lines are the accuracy of the improved VGG-19 model and the improved Inception V3 model based on DCGAN data augmentation proposed in this paper. We can see that the improved VGG-19 model proposed in this paper has the highest accuracy. After data augmentation, the accuracy of the improved VGG-19 model is higher than the improved Inception V3 model.

Table 7. Evaluation of improved Inception V3 model before and after data augmentation

Evaluation of improved inception V3 before data augmentation				Evaluation of improved inception V3 after DCGAN			
Type	Precision	Recall	Specificity	Type	Precision	Recall	Specificity
G	94.25%	86.19%	95.42%	G	93.85%	91.19%	94.80%
M	72.97%	83.90%	90.66%	M	79.48%	85.42%	93.37%
P	83.62%	85.05%	92.74%	P	89.43%	88.22%	95.46%

Table 8. Comparison with other methods using the figshare dataset

Method	Accuracy (%)
Afshar [5]	86.56
Zia [6]	85.69
Afshar [20]	90.89
Cheng [14]	91.28
Improved inception V3	88.96
Improved VGG-19	91.73

5 Conclusion

In this paper, we proposed a 3-class brain tumor classification system based on data augmentation and improved CNN framework, our system has high accuracy and specificity, and the system can be applied to mobile medical equipment for the positive diagnosis. In this paper, we solved the problem of insufficient data samples in medical image classification tasks and analyzed the impact of data augmentation based on image processing technique and data augmentation based on DCGAN on model performance. We improved the VGG-19 model and the Inception V3 model, used Adam algorithm to optimize two CNN models, and discussed the impact of different parameters on the accuracy of the model. Comparing the accuracy of the two models with the augmented dataset before and after improvement, it showed that the improvement of the two models is effective. Comparing with other methods, the classification accuracy of the proposed system was the highest, and our algorithm has better performance, which provides the possibility for migration to mobile medical equipment and can improve the diagnostic performance of mobile medical equipment. But this paper still needs several improvements: 1) Too many data samples may cause the model to overfit. 2) When using transfer learning, the impact of different size datasets on model performance is not discussed. 3) Did not solve the problem of data imbalance.

Acknowledgments. This work is supported by National Natural Science Foundation of China (Grant nos. 81772009), Science and Technology Project of Henan Province Science and Technology Department (Grant nos. 182102310162), Key scientific research projects of Henan universities in 2017 (Grant nos. 17A520014), 2017 Henan Province Science and Technology Project (Grant nos. 172102310496), 2019 Zhengzhou University offline excellent course construction project (Grant nos. 2019XXJPKC023), Research on Video Tracking Moving Object Modeling (Grant nos. 2019ZDGGJS029).

References

1. Laukamp, K.R., Thiele, F., Shakirin, G., et al.: Fully automated detection and segmentation of meningiomas using deep learning on routine multiparametric MRI. Eur. Radiol. **29**(1), 124–132 (2019)
2. Kamnitsas, K., Ledig, C., Newcombe, V.F.J., et al.: Efficient multi-scale 3D CNN with fully connected CRF for accurate brain lesion segmentation. Med. Image Anal. **36**, 61–78 (2017)
3. Kharrat, A., Gasmi, K., Messaoud, M.B., et al.: A hybrid approach for automatic classification of brain MRI using genetic algorithm and support vector machine. Leonardo J. Sci. **17**(1), 71–82 (2010)
4. Zhu, H., Fang, Q., He, H., et al.: Automatic prediction of meningioma grade image based on data amplification and improved convolutional neural network. Comput. Math. Methods Med. **2019** (2019)
5. Afshar, P., Mohammadi, A., Plataniotis, K.N.: Brain tumor type classification via capsule networks. In: 2018 25th IEEE International Conference on Image Processing (ICIP), pp. 3129–3133. IEEE (2018)
6. Zia, R., Akhtar, P., Aziz, A.: A new rectangular window based image cropping method for generalization of brain neoplasm classification systems. Int. J. Imaging Syst. Technol. **28**(3), 153–162 (2018)
7. Swati, Z.N.K., Zhao, Q., Kabir, M., et al.: Brain tumor classification for MR images using transfer learning and fine-tuning. Comput. Med. Imaging Graph. **75**, 34–46 (2019)
8. Deepak, S., Ameer, P.M.: Brain tumor classification using deep CNN features via transfer learning. Comput. Biol. Med. **111**, 103345 (2019)
9. Goodfellow, I.: NIPS 2016 tutorial: generative adversarial networks. arXiv preprint arXiv: 1701.00160 (2016)
10. Chang, K., Bai, H.X., Zhou, H., et al.: Residual convolutional neural network for the determination of IDH status in low-and high-grade gliomas from MR imaging. Clin. Cancer Res. **24**(5), 1073–1081 (2018)
11. Han, C.: Infinite brain MR images: PGGAN-based data augmentation for tumor detection. In: Esposito, A., Faundez-Zanuy, M., Morabito, F.C., Pasero, E. (eds.) Neural Approaches to Dynamics of Signal Exchanges. SIST, vol. 151, pp. 291–303. Springer, Singapore (2020). https://doi.org/10.1007/978-981-13-8950-4_27
12. Radford, A., Metz, L., Chintala, S.: Unsupervised representation learning with deep convolutional generative adversarial networks. arXiv preprint arXiv:1511.06434 (2015)
13. Kingma, D.P., Ba, J.: Adam: a method for stochastic optimization. arXiv preprint arXiv:1412. 6980 (2014)
14. Cheng, J., Huang, W., Cao, S., et al.: Enhanced performance of brain tumor classification via tumor region augmentation and partition. PloS ONE **10**(10) (2015)
15. Otsu, N.: A threshold selection method from gray-level histograms. IEEE Trans. Syst. Man Cybern. **9**(1), 62–66 (1979)
16. Xue, D.X., Zhang, R., Feng, H., et al.: CNN-SVM for microvascular morphological type recognition with data augmentation. J. Med. Biol. Eng. **36**(6), 755–764 (2016)
17. Simonyan, K., Zisserman, A.: Very deep convolutional networks for large-scale image recognition. arXiv preprint arXiv:1409.1556 (2014)

18. Szegedy, C., Vanhoucke, V., Ioffe, S., et al.: Rethinking the inception architecture for computer vision. In: Proceedings of the IEEE Conference on Computer Vision and Pattern Recognition, pp. 2818–2826 (2016)
19. Hinton, G.E., Salakhutdinov, R.R.: Reducing the dimensionality of data with neural networks. Science **313**(5786), 504–507 (2006)
20. Afshar, P., Plataniotis, K.N., Mohammadi, A.: Capsule networks for brain tumor classification based on MRI images and coarse tumor boundaries. In: ICASSP 2019 – 2019 IEEE International Conference on Acoustics, Speech and Signal Processing (ICASSP), pp. 1368–1372. IEEE (2019)

Evaluating the Effectiveness of Inhaler Use Among COPD Patients via Recording and Processing Cough and Breath Sounds from Smartphones

Anthony Windmon[1]([✉]), Sriram Chellappan[1], and Ponrathi R. Athilingam[2]

[1] Department of Computer Science and Engineering, University of South Florida, Tampa, FL 33620, USA
{awindmon,sriramc}@usf.edu
[2] College of Nursing, University of South Florida, Tampa, FL 33612, USA
pathilin@usf.edu

Abstract. Chronic Obstructive Pulmonary Disease (COPD) is a major health concern for elders today. Chronic cough and wheezing, which occur in the lungs as a result of mucus buildup are the main symptoms of COPD. COPD patients are advised to regularly medicate themselves via an inhaler, which delivers medicine to the lungs to break down mucus and relieve wheezing. Unfortunately, many patients do not use their inhaler devices correctly, resulting in no improvement of COPD symptoms, and worsened health. In this paper, we design machine learning (Support Vector Machine) algorithms operating on Mel-frequency Cepstral Coefficients of cough and breath sounds of patients (recorded via smartphones before and after inhaler usage) to detect the effectiveness of inhaler usage. Using a cohort of 55 clinically diagnosed COPD patients, spread across both genders, we evaluate our system from multiple metrics, including Precision, Recall, Sensitivity and Specificity. Our system achieved accuracies close to 80% in detecting effectiveness of inhaler usage. Our proposed system can aid COPD patients in improved selfcare routines, and also reduce the rate of re-hospitalizations caused by exacerbated symptoms.

Keywords: COPD · Lungs · Health · Cough · Breath · Machine learning · Smartphones · Aging

1 Introduction

Chronic Obstructive Pulmonary Disease (COPD) is the fourth leading cause of death worldwide, and is estimated to become the third by 2020 [1]. The prevalence of COPD in the US alone is 24 million patients today. However it is estimated that the number is actually much higher, due to millions not diagnosed yet, but who are living with impaired lung functions [2]. The most common cause of COPD is smoking, accounting for 85%

© ICST Institute for Computer Sciences, Social Informatics and Telecommunications Engineering 2020
Published by Springer Nature Switzerland AG 2020. All Rights Reserved
J. Liu et al. (Eds.): MobiCASE 2020, LNICST 341, pp. 102–120, 2020.
https://doi.org/10.1007/978-3-030-64214-3_7

of cases, while occupational smoke/dust and genetic factors are responsible for COPD in the remaining 15% of the population today [2].

Chronic cough and wheezing from the lungs are main symptoms of COPD, due to excess mucus production. Pharmacological therapy for COPD includes regular self-use of an inhaler (to deliver medicine directly to the lungs to breakdown mucus), and is validated in several clinical trials [3, 4]. However, it is a fact that a significant percentage of patients engage in sub-optimal inhaler techniques during self-care [5], which, as a consequence does not breakdown mucus enough, leading to worsened symptoms/health, and sometimes re-hospitalizations.

Our Contributions: In this paper, we propose an in-home smartphone based system to enable a COPD patient to determine the effectiveness of inhaler use via processing cough and breath sounds. To do so, we recruited a cohort of 55 clinically diagnosed COPD patients, spread across both genders[1]. Each subject was asked to cough and take deep breaths before inhaler use (to detect presence of mucus) and after correct inhaler use (to detect break-up of mucus and symptoms improvement). All data was recorded via a smartphone. After removing noise, three experts (one of them, our third co-author) listened to each audio segment recorded, to classify the cough and breath sounds as symptomatic of COPD (i.e., excess mucus build-up) or otherwise (i.e., mucus breakup due to correct inhaler use). After appropriate pre-processing, a total of 430 s of cough audio, and 1161 s of breath audio were obtained, evenly spread before and after inhaler use.

From this audio dataset, we then extracted Mel-frequency Cepstral Coefficients (MFCC) for post-processing. Very briefly, the mel-frequency cepstrum (MFC) is a representation of the short-term power spectrum of an audio signal, based on a linear cosine transform of a log power spectrum on a nonlinear mel scale of frequency. The Mel-frequency cepstral coefficients (MFCC) are those that together make up an MFC. MFCC provides a robust feature set for our classification problem (i.e., assess improvement in cough and breath due to inhaler use), as this feature performs best at capturing the spectral envelope of cough and breath sounds. As we present later in the paper, the spectral envelope is a critical component in audio signal processing that best captures features unique to sounds like cough and breath.

We then designed a spectrum of machine learning algorithms to process the MFCC extracted in order to classify cough and breath of COPD patients. Specifically, we want to discern those cough and breath sounds that indicate absence of mucus (with correct inhaler use) compared to sounds that indicate presence of mucus. It is easy to see that if we are successful in achieving our goal, quick feedback can be given to patients indicating either a) they are correctly using the inhaler; or b) they are incorrectly using the inhaler (while also directing them to tutorials on correct inhaler use).

To address our goal, we found that a Support Vector Machine (SVM) algorithm performed the best among k-Nearest Neighbors, Random Forests, and Logistic Regression in terms of standard metrics like Precision, Recall, Sensitivity and Specificity. Our overall accuracies were close to 80% for both cough and breath. We believe that our

---

[1] The study was approved by the University of South Florida's Institutional Review Board (IRB). **IRB Reference Number: Pro00035013.**

paper is the first to actually design an in-home smartphone based system to record and process cough & breath to evaluate the effectiveness of inhaler use. We expect our system to have significant value to educate patients, improve health outcomes and reduce re-hospitalization rates in relation to COPD.

The remainder of this paper is organized as follows: Sect. 2 covers related work. Section 3 explains our data collection procedures. Section 4 elaborates on data processing, and Sect. 5 details our feature extraction and classification techniques. Section 6 presents evaluation results, and, finally, we conclude and discuss future works in Sect. 7.

2 Related Work

In this section, we elaborate upon important related work within the scope of this paper.

Analyzing Cough for Healthcare: There are a number of studies in the recent past that process cough for healthcare. Researchers collected cough sounds from 38 subjects, 17 with tuberculosis and 21 without, to develop an algorithm capable of detecting symptoms of Tuberculosis [6]. Cough samples were recorded, voluntarily, from both infected and uninfected patients, using a Tascam DR44WL hand-held audio recording device with a 44100 Hz sampling rate. The features processed were log spectral energies and MFCC (similar to our system in this paper), and the classifier combines decision trees and logistic regression methods. For this problem, the authors achieved an 82% accuracy, 95% sensitivity, 72% specificity, and an area under the curve (AUC) score of 0.95 [6].

There are also systems that process audio signals to *only* detect cough. Related works in this space are [7] and [8], where data is collected using smartphones and close to 1000 s of cough served as the training dataset. Using machine learning techniques like k-NN and SVM, accuracies close to 90% were achieved in classification. In another paper, a slightly more complex problem - namely detection of respiratory events (cough, sneezing, throat-clearing and sniffling) is addressed [9]. Data pertaining to these events was recorded from a cohort of 16 subjects using a smartphone, unobtrusively for six weeks. This technique is a multi-layered SVM approach that processes several time and frequency domain features. In this paper, researchers achieved an accuracy of 82% for detection of respiratory events, and 99.1% for detection of non-respiratory events [9]. More recently, and with advances in deep learning via neural networks, there are some works that design convolutional neural network (CNN) based techniques in the domain of cough detection. Work in this domain include [10, 11] and [12]. In these papers though, the number of subjects recruited was relatively small (ranging from only 9 to 14), and accuracies close to 95% were achieved.

While these are all important related work, we point out that the mere detection of cough was the problem of interest here, and not finer grained classification of cough for a specific health condition (i.e. COPD) as we do in this paper.

Breath Analysis for Healthcare: Processing breath sounds is important for healthcare. Machine Learning algorithms have been designed and implemented to analyze breathing techniques to differentiate patients with lung cancer from healthy patients, and from a mixed group of patients with other lung diseases (i.e., COPD, asthma, pneumonia, etc.)

in works like [13]. Also, systems to detect various phases on breathing (without a specific disease context) have been developed in works like [14]. Other related work in the space of breath detection is [15], where algorithms are devised to measure lung function, including exacerbation, by processing breath sounds via a spirometer connected to a smart-phone. There are also many papers related to processing breath sounds for sleep and exercise detection, which we do not elaborate here due to space limitations.

Our Prior Work: We have done prior work in the space of designing AI techniques to process cough. In our first paper [16], we designed a classification system to discriminate cough indicative of COPD symptoms, from cough collected from healthy subjects that do not have COPD, wherein the cough was recorded via smartphones. In our second paper [17], we expanded our system by enabling it detect COPD and CHF (Congestive Heart Failure) from normal cough. In this paper, we have similar, but still orthogonal goals, in that we are now attempting to detect specifically the effectiveness of inhaler use by monitoring and analyzing cough and breath sounds before and after its use. To the best of our knowledge, this is a unique problem not yet attempted in the literature, but one that has big impact.

A Note on Sample Sizes: We wish to point out that the process of collecting data (elaborated in the next section) was difficult. We could only recruit 55 patients with COPD that met our criteria for recruitment (including age, gender, mental health conditions, approved for inhaler use etc.), and it was a nine month effort. Many patients did not consent to our study, and it is normal to do so. As such, in this paper, we do not attempt deep learning (i.e., featureless) techniques, due to non-availability of truly big data. But we are confident that our machine learning techniques in this paper are rigorous. Attempting deep learning techniques is part of our future work with much more data collected.

3   Data Collection

3.1   Recruitment of Subjects with COPD

During Spring and Summer 2019, we collaborated with respiratory therapists at Tampa General Hospital in Downtown Tampa, FL. With their assistance, we identified 55 (34 Female and 21 Male) clinically diagnosed COPD patients. Each subject was asked to sign an Institutional Review Board (IRB) approved consent form, indicating their willingness to participate in our study. Additionally, subjects were asked to provide their demographic information (i.e. age, gender, martial status, etc.), documented in Table 1, and to complete the COPD ABC and Leicester Cough Questionnaires. The COPD ABC Questionnaire measures the burden of COPD [22], and the Leicester Cough Questionnaire assesses the impact of cough on various aspects of life (i.e., personal, professional, etc.) [23].

3.2   Our Procedure for Recording Cough and Breath Sounds

Cough and breath sounds were recorded using a custom application developed by the authors. This application was installed onto a Motorola Moto E SmartPhone device, containing Android version 4.4.4 KitKat, recording at a sampling rate of 44100 Hz and bit rate of 16 (per second). This sample and bit rate are standard [34] and recommended [55].

Table 1. Demographic information of subjects.

Description	Category		Data
		(N=55)	62.09 ±
	Mean		
	Std. Deviation	12.54	
	Female	34 (61%)	
	Male	21 (39%)	
	Married	16 (29.6%)	
	Never Married	11 (18.5%)	
	Windowed, Divorced or Separated		28 (51.9%)
Education:	<High School		11) (18.5%
	High School or GED	16 (29.6%)	
	Some College	12 (22.2%)	
	Degree or Professional	16 (29.6%)	
Mean of COPD	Current Smoker	9 (16.7%)	29.5
Score: Burden	Score on COPD ABC Questionnaire		5.11 ±
Mean COPD	Score on Leicester Cough		8.57 ±
Severity Score:	Questionnaire		3.59

We collected *four* samples from each subject: (1) cough and (2) breath sounds before inhaler use, and (3) cough and (4) breath sounds after inhaler use. To collect cough, each subject would simply cough into the phone's microphone via our app. However, to collect breath sounds (i.e., wheezing), we developed a custom recorder using the diaphragm of an actual stethoscope. Connected it to is an Audio-Technica ATR-3350IS Omnidirectional Condenser Lavalier microphone, which records breath sound waves as audio files. The microphone, shown in Fig. 1, connects to our smart-phone via wire and is available in the market.

The process of collecting data required care. Initially, *before* each subject was administered their inhaler medication, we recorded a sample of their cough. To do so, each subject was asked to cough directly into the microphone of our smartphone in which the sound was recorded. Recording breath required a series of steps. First, we started off using an actual stethoscope to manually listen to each subject's wheezing sounds in their lungs, which indicate quality of breath sounds. The objective was to locate the clearest wheezing sounds being projected. Most often, wheezing sounds are best heard on four different areas of the subject's front, mid-to-lower chest, where the lungs are located. Additionally, wheezing sounds can be heard on eight different areas of the subject's mid-to-lower back. The best areas of auscultation, as depicted in Fig. 2, vary by subject. Once we identify the best location where wheezing was heard, we used our

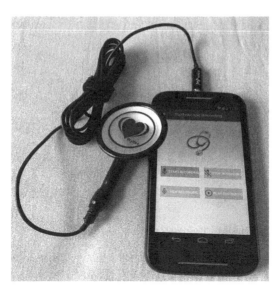

Fig. 1. The Audio-Technica ATR-3350IS omnidirectional condenser lavalier microphone (with heart shape on surface) connected to the Motorola Moto E smart-phone device to collect breath sounds from COPD patients. The custom mobile application is shown on the screen of the smart-phone.

stethoscope, Audio-Technica microphone and mobile application to record and store the clearest wheezing sound from that location for that subject. All recorded data was appropriately labeled (without any identifiable information).

In the second round of recording, the subject was required to take their inhaler medication, which was correctly administered by a respiratory therapist. About five minutes *after* inhaler use, the exact same cough and breath recording process was executed, and all data was labeled. This process was repeated for all subjects. Since all patients used the inhaler correctly, their cough and breath sounds sounded differently due to mucus break down after inhaler use (which can be gleaned by a trained ear). Automating the classification via this ground-truth data is our problem.

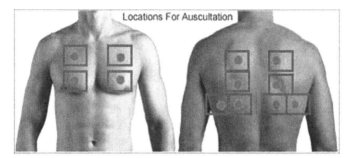

Fig. 2. Locations for auscultation, where wheezing sounds are heard best. Front, mid-to-lower chest (left), mid-to-lower back (right).

Table 2. Number of seconds (sec.) for cough and breath classes, before (BP) and after (AP) data pre-processing.

Class	Sec. (BP)	Sec. (AP)
Cough before inhaler use	289 s	219 s
Breath before inhaler use	627 s	579 s
Cough after inhaler use	261 s	211 s
Breath after inhaler use	632 s	582 s

4   Data Pre-processing

4.1   Cough and Breath Data Before Pre-processing

In our data collection procedures, cough sounds, on an average, lasted about 7 s each. Breath sounds, on the contrary, were collected for approximately one minute each, consisting of 5–7 deep breaths. Using this data, we developed four different classes: cough before inhaler use, breath before inhaler use, cough after inhaler use and breath after inhaler use. As shown in Table 2, the second column contains the duration of raw data before any pre-processing (denoted as BP). Data in the third column is after pre-processing (denoted as AP), and is discussed next.

4.2   Pause and Noise Removal

First, we identified unintentional pauses in our audio files post recording, and cut them using an audio cutting application [17]. Next, we applied a band-pass filter to lower background ambient noise caught in our cough data (i.e., noise due to surrounding conversations, medical equipment, etc.), with cut-off frequencies as 300 Hz and 1200 Hz because cough lies in-between those frequencies [20, 21]. There was little to no noise in our breath data, because the stethoscope was placed tightly on the subject's skin for recording.

4.3   One Second Windowing Algorithm

The next issue is accurate ground-truthing of data collected. This is a little tricky. Even if a patient is clinically diagnosed with COPD and chronic cough, it is still not the case that the entire cough episode indicates symptoms of COPD. It often happens that only a certain subset of their entire cough episode indicates COPD symptoms. Naturally, with proper inhaler use, subsequent cough segments will be completely devoid of COPD symptoms. Fortunately, the third co-author of this paper has decades of experience working with COPD patients, and she indicated that a one second segment of cough can indicate presence of COPD symptoms to a trained human ear. The same is true for breath also. As such, we split our entire dataset of cough and breath (after pause and noise removal) from all patients into one second segments and our third co-author, joined by other experienced nurses, listened to each cough and breath segment to tag it as indicative of concerning COPD symptoms (that necessitated inhaler intervention for that episode of cough/breath) or very mild/no discernible COPD symptoms (that does not necessitate any further inhaler use for that episode).

As a result of these procedures, we generated a new dataset containing 219 s of cough and 579 s of breath audio that indicated COPD symptoms necessitating inhaler intervention; and 211 s of cough and 582 s of breath audio that demonstrated improved COPD conditions not warranting any further inhaler use for that episode. This data is shown in the third column in Table 2. This dataset is what we train and validate our machine learning algorithms on.

5   Feature Extraction & Classification Algorithms

We now elaborate on extraction of Mel Frequency Cepstral Coefficients (MFCC) from our dataset as features, and our Support Vector Machine based algorithm for classification.

5.1   Mel Frequency Cepstral Coefficients

For sound recognition systems, a primary goal is to classify a sound (i.e., speech, singing, breath, cough, etc.) as produced by a human. Human sounds are produced via the larynx (voice box) and vibrations of vocal cords. This sound is then filtered by their vocal tract, which determines how the sound produced is, *both*, shaped and ejected from the mouth. The vocal tracts for a human consists of the lips, nose, tongue, teeth and throat areas [43], as depicted in Fig. 3. The corresponding shape of the sound is defined within the envelope of the short time power spectrum, which estimates loudness and timbre, also shown in Fig. 3. The MFCC is the strongest audio feature capable of accurately defining that envelope [26, 35], which inturn serves as robust features to classify sounds like cough and breath, since the shape of the vocal tract defines how these sounds emanate [39–41]. The MFCC accomplishes this by generating Cepstral Coefficients. The calculation to generate the Cepstral Coefficients, depicted for each of our four classes in Fig. 4, is explained below and mapped out in Fig. 5.

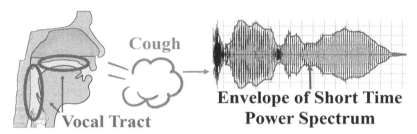

Fig. 3. Sound type (i.e. sound of cough or breath) is produced by the shape of the vocal tract. This shape is defined within the envelope of the Short Time Power Spectrum, which is best characterized by cepstral coefficients of the MFCC.

Computing MFCC Features: First, each audio (i.e., cough or breath) signal h is split into a small number of frames of duration 20 ms (ms). If the frame is shorter we do not have enough samples to get a reliable spectral estimate. If it is longer, the signal changes too much throughout the frame, making it highly non-stationary [26]. Since, our sampling rate to record the cough and breath audio signals is 44100 Hz, the frame length s of each audio signal is now $0.020 * 44100 = 882$ samples. Next, from each frame, we extract one set of 13 MFCC coefficients.

To do so, we denote our pre-framed audio signal as $h(s)$, and our framed audio signal as $h_i(s)$. Then, we calculate the Discrete Fourier Transform (DFT) $D_i(k)$ of each ith frame as,

$$D_i(k) = \sum_{s=0}^{S-1} h_i(s)w(s)x(s)e^{\frac{2\pi ks}{S}}. \tag{1}$$

Here, $w(s)$ represents the length of the hamming window function, where $w(s) = 0.54 - 0.46 \cos(\pi s/S)$. S is the length of the discrete-time signal $x(s)$, which represents the quantity of signals in s. Also, k is the sampling frequency of the DFT, where $k = 0,1...S - 1$.

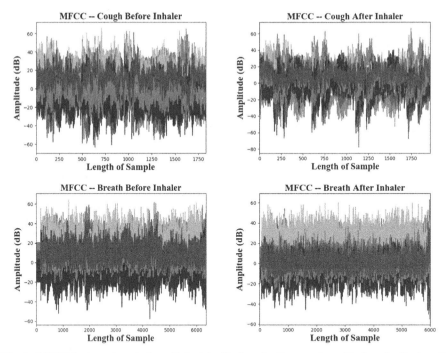

Fig. 4. MFCC features (cepstral coefficients) of before & after cough, and before & after breath samples, which are represented in the shape of the envelope of the short time power spectrum in Fig. 3. These images represent the lower 13 MFCC coefficients, which contain the highest quantity of information about the overall spectral shape produced by cough and breath sounds. For cough, there's a difference in amplitude due to reduction of mucus build-up after inhaler use. For breath, there's a difference in amplitude consistency, also due to reduction in mucus build-up after inhaler use. This figure is best viewed in color.

We then compute the Periodogram Estimate $M_i(k)$ as,

$$M_i(k) = \frac{1}{S}|D_i(k)|^2, \tag{2}$$

to identify which frequencies are present in each frame, and decipher cough and breath sound frequencies analogous to the human ear [26].

Next, to produce the Mel-Filter Bank, we applied a Triangular Filter, depicted in Fig. 6 [27]. The Triangular Filter, roughly, captures energy within the spectral envelope of a frequency bin. In other words, the filter provides an estimate of the given audio sample's spectral envelope shape. The Triangular Filter is applied on a Mel-Scale to the power spectrum. The Mel-Scale imitates the linear frequency of the human ear's perception to sound.

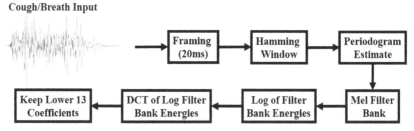

Fig. 5. The multitude of steps required to calculate the mel frequency cepstral coefficients (MFCC) for both cough and breath sounds.

The Mel-Filter Bank contains 26 vectors, and 257 coefficients. We multiple each Filter Bank by the power spectrum, calculated using Eq. 2, then add the coefficients. This results in 26 numbers, representing the amount of energy in each Filter Bank. The logarithm of these 26 numbers is calculated, which imitates what is heard by a human ear.

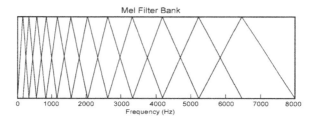

Fig. 6. Triangular filter applied to produce the mel filter bank [27].

Finally, we take the Discrete Cosine Transform (DCT) of the 26 log numbers. This results in 26 cepstral coefficients. We only keep the lower 13 coefficient, as these coefficients contain the strongest quality of information about the spectral envelope's shape [26]. We discard the higher coefficients because they represent fast changes in the

Mel-Filter Bank energies, which decrease cough and breath recognition performance. We see a small, but noticeable, increase in performance by dropping the higher coefficients.

After applying the DCT, the MFCC can be expressed as,

$$C_m = \sum_{k=1}^{K} (logD_i(k)) \left[m \left(k - \frac{1}{2} \right) \frac{\pi}{K} \right], \tag{3}$$

where C_m denotes the MFCC, $m = 1,2 \ldots 13$, which equates to the 13 MFCC coefficients, and $K = 44100$ which is the sampling rate of our recording device.

5.2 Justification for MFCC Audio Features

As of today, health-care professionals listen to the sound of a patient's cough and wheezing to determine the presence and severity of COPD [50]. The point of our study is to automate this process using machine learning processed on cough and breath. To do so, we need to capture the spectral envelope of cough and wheezing sounds. In the literature, it has been demonstrated various times that the MFCC audio features are most capable of capturing the spectral envelope [26, 43]. The MFCC does so by converting sounds to a Mel-Scale which analyses those sounds at frequencies that humans speak, and are capable of hearing. The MFCC is an ideal feature for our problem for the following reasons: (1) MFCC uses the mel-scale to analyze sound in a manner similar to humans [26]; (2) It has been successfully used in several similar applications related to cough [16, 20, 45, 46], breath [47], wheezing [48, 49], music [51] and speech recognition systems [52]; (3) MFCC is considered a classic front-end algorithm capable of significant and accurate performance in sound and speech recognition systems [43, 44]. Lastly, there are several studies that suggest a significant relationship between the MFCC audio feature and Support Vector Machine classification combination (which were both utilized in this study), when used in the domains of cough [12, 52, 53] and breath [48, 54] analysis.

Other audio features, like Spectral Centroid, Spectral Flatness or Spectral Flux, could have been used to solve this problem. However, these features do not capture the spectral envelope's perception as the MFCC does. Thus, we did not incorporate those features into this study. Furthermore, using a large number of features on our classification models can cause overfitting problems and also increased overhead. Hence, we stick with MFCC features alone for our problem, and are confident about our decision to do so.

5.3 Support Vector Machine

Based on MFCC features presented above, we briefly present our Support Vector Machine (SVM) based algorithm for classification, which performed the best among other techniques. Broadly speaking, SVM classifiers aim to find the best hyperplane between two classes. A hyperplane is a line which linearly separates the data points between the classes. In SVM, a hyperplane is considered "best" when it produces the largest margin between two classes. SVM uses Support Vectors, which are the classes' data points closest to the hyperplane, to calculate margin maximization [28]. Recall again that our problem is to identify improvement in symptoms before and after inhaler usage using cough and breath data in Table 2. We design two separate SVM classifiers to do so – one to process cough and the other to process breath.

Our classifiers for cough and breath were developed using the scikit-learn machine learning software, built into python programming language. The parameters for our SVM classification models, which produced best results were as follows:

Cough: degree of the Radial Basis Function (RBF) kernel function is 7, the cache size is 120, the random state is 4 and the kernel is linear, regularization = 1.0, tolerance for stopping criterion = 1e−3, class weight = balance, decision function shape = one-vs-rest, maximum number of iterations = −1 (default), gamma = "scale" and shrinking = True.

Breath: degree of the Radial Basis Function (RBF) kernel function = 5, the cache size = 215, the random state = 4, the kernel is linear, regularization = 1.0, tolerance for stopping criterion = 1e−3, class weight = balance, decision function shape = one-vs-rest, maximum number of iterations = −1 (default), gamma = "scale" and shrinking = True.

The RBF kernel was selected and works best for our study because (1) parts of our data has overlap making it difficult for SVM to find the right hyperplane to separate the data; (2) RBF provides better discriminative ability in a much higher dimensional subspace [28]; (3) RBF provided a much faster classification time in comparison to the Polynomial kernel.

6 Results

Classification results, presented below were measured using the following cross validation methods: 10-Fold (10-FCV) and Leave-One-Out (LOOCV). Metrics are Specificity, Sensitivity, Precision, Recall and F1-Score. Results of our SVM classifier, as well as other machine learning algorithms are recorded in Tables 3, 4, 5 and 6.

Cough: For our cough classification scheme, testing the cough before inhaler and cough after inhaler classes, we achieved the following results. Employing 10-Fold Cross Validation, we averaged an accuracy of 79.00%, precision of 81.00, recall of 81.00%, sensitivity of 84.52%, specificity 77.61% and a F1-score of 81.00%. Employing Leave-One-Out Cross Validation, we averaged an accuracy of 80.69%, precision of 80.00, recall of 80.00%, sensitivity of 83.45%, specificity 78.02% and a F1-score of 80.00%.

Breath: For classification, testing the breath before inhaler and breath after inhaler classes, we achieved the following results. Employing 10-Fold Cross Validation, we averaged an accuracy of 84.49%, precision of 84.00, recall of 83.00%, sensitivity of 83.00%, specificity 93.30% and a F1-score of 82.00%. Employing Leave-One-Out Cross Validation, we averaged an accuracy of 84.32%, precision of 83.00, recall of 83.00%, sensitivity of 82.00%, specificity 91.49% and a F1-score of 80.00%.

6.1 Comparison of Results Using Other Algorithms

Tables 3, 4, 5 and 6 show classification results using several popular machine learning approaches (i.e. k-Nearest Neighbors, Random Forests, Logistic Regression and Multilayer Perceptron). As shown, Support Vector Machine (SVM) provided the best results for the majority of metrics. This is because SVM works best for binary classification problems, and also works well with linearly separable data [28], such as our data. Figure 7 illustrates the Receiver Operating Characteristic (ROC) curves for classification performance based on cough and breath data, with SVM performing the best with Area under the Curve (AUC) scores close to 94% for cough and 93% for breath, when applying 10-Fold Cross Validation. When applying Leave-One-Out Cross Validation, the AUC scores are 87.5% for cough and 88.4% for breath.

6.2 Complexity of Execution

Our Support Vector Machine model, was compiled on a Windows 10 Dell PC, containing an Intel(R) Core(TM) $i7$-5600U processor, 2.60 GHz with 16 GB RAM. All processing (i.e. audio cutting/filtering, feature extraction and classification) was done with this system. MATLAB R2017b software was used for feature extraction, audio filtering and

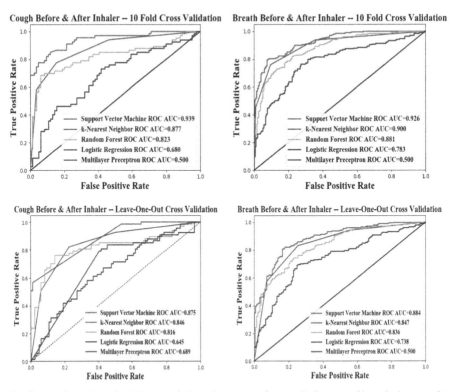

Fig. 7. Receiver Operating Characteristic (ROC) curves for cough classes and breath classes using 10-fold (top) and leave-one-out cross validation (bottom).

Table 3. Classification of cough based on 10-fold cross validation.

Algorithm	Accuracy precision recall sensitivity specificity F1-score				
Support vector machine 79.00%	81.00% 81.00%	84.52%	77.61%	81.00%
k-Nearest neighbors 77.00%	81.00% 81.00%	83.33%	77.61%	81.00%
Random forests 65.00%	76.00% 51.00%	92.00%	77.60%	48.00%
Logistic regression 66.00%	73.00% 73.00%	77.27%	69.04%	73.00%
Multilayer perceptron 62.00%	20.00% 44.00%	74.32%	54.66%	27.00%

Table 4. Classification of cough based on leave-one-out cross validation.

Algorithm	Accuracy precision recall sensitivity specificity F1-score				
Support vector machine 80.69%	80.00% 80.00%	83.45%	78.02%	80.00%
k-Nearest neighbors 79.06%	82.00% 82.00%	82.57%	79.00%	81.00%
Random forests 62.08%	74.00% 53.00%	89.00%	79.45%	46.00%
Logistic regression 67.45%	71.00% 71.00%	75.54%	68.65%	71.00%
Multilayer perceptron 58.85%	22.00% 43.00%	73.08%	52.27%	28.00%

Table 5. Classification of breath based on 10-fold cross validation.

Algorithm	Accuracy precision recall sensitivity specificity F1-score				
Support vector machine 84.49%	84.00% 83.00%	83.00%	93.30%	82.00%
k-Nearest neighbors 84.00%	84.00% 83.00%	76.00%	90.40%	83.00%
Random forests 80.00%	82.00% 81.00%	74.16%	87.87%	81.00%
Logistic regression 70.00%	72.00% 71.00%	77.27%	65.04%	71.00%
Multilayer perceptron 50.00%	23.00% 48.00%	46.56%	38.66%	31.00%

Table 6. Classification of breath based on leave-one-out cross validation.

Algorithm	Accuracy precision recall sensitivity specificity F1-score				
Support vector machine	84.32%	83.00% 83.00%	82.00%	91.49%	80.00%
k-Nearest neighbors	83.89%	82.00% 80.00%	73.00%	87.32%	81.00%
Random forests	82.25%	83.00% 80.00%	75.22%	86.32%	81.00%
Logistic regression	70.70%	73.00% 70.00%	78.01%	66.55%	72.00%
Multilayer perceptron	47.71%	25.00% 46.00%	44.52%	40.33%	34.00%

dataset construction. The custom mobile application used to record cough and breath samples, consumed a memory 4.55 MB on Motorola Moto E Smart-Phone containing 270 MB. Python 3.6 Sklearn libraries were used to develop all classifiers. The classification time during execution on the Windows 10 Dell PC was less than a minute. Ensuring all processing and classification to happen within the smartphone itself is part of future work.

7 Conclusions and Future Work

In this paper, we design a system capable of discerning improvements in cough and breath symptoms for COPD patients as a result of correct inhaler use (due to reduced mucus build-up). Data from 55 patients was used to developed our model. Our ultimate goal is to improve health of the individual and reduce costly re-hospitalizations, by detecting and recommending correct health procedures in-home. Our results are very favorable. We also believe that our work is the first to address the issue of designing algorithms to automatically evaluate the effectiveness of prescribed interventions (in this case, inhalers) for COPD.

In the future, we propose to integrate our algorithm as an AI based in-home care system for patients with COPD. With immediate feedback upon symptom monitoring, patients could be better notified of when to use an inhaler, and also offered tutorials on correct inhaler use which can definitely reduce re-hospitalizations. This system will also include the ability to inform patients when their cough symptoms have exacerbated enough for them to seek medical attention. Integrating these AI designs into current COPD care products in the market like [29–32, 37] is our strategy.

Also, we are aiming to conduct a longer study to design gender based algorithms, since clinical studies suggests that women experience significantly harsher COPD symptoms than men, throughout their lifespan [36]. While in this paper, we processed cough and breath separately, we are also looking into integrate both sounds and related features together to detect correctness of inhaler usage. Conducting longitudinal experiments to design personalized models for each patient is also part of our future work.

Acknowledgment. Support for Anthony Windmon was provided by the Florida Education Fund's McKnight Doctoral Fellowship Program. We thank registered nurses and scientists, respiratory therapists and participating patients at Tampa General Hospital for their support and cooperation.

References

1. Global Initiative for Chronic Obstructive Lung Disease (GOLD): GOLD 2020 global strategy for the diagnosis, management, and prevention of chronic obstructive pulmonary disease, 2020 report. https://goldcopd.org/wpcontent/uploads/2019/12/GOLD-2020
2. Center for Disease Control and Prevention (CDC): Chronic Obstructive Pulmonary Disease: Data and Statistics, May 2018. https://www.cdc.gov/copd/data.html
3. Barnes, N., Calverley, P.M., Kaplan, A., Rabe, K.F.: Chronic obstructive pulmonary disease and exacerbations: clinician insights from the global Hidden Depths of COPD survey. Curr. Med. Res. Opin. **304**, 667–684 (2014). https://doi.org/10.1185/03007995.2013.867842

4. Dolovich, M.B., et al.: Device selection and outcomes of aerosol therapy: evidence-based guidelines: American College of Chest Physicians/American College of Asthma, Allergy, and Immunology. Chest **127**(1), 335–371 (2005). https://doi.org/10.1378/chest.127.1.335

5. Melani, A.S., et al.: Inhaler mishandling remains common in real life and is associated with reduced disease control. Respir. Med. **105**(6), 930–938 (2011). https://doi.org/10.1016/j.rmed.2011.01.005

6. Botha, G.H.R., et al.: Detection of tuberculosis by automatic cough sound analysis. Physiol. Meas. **39**(4), 045005 (2018). https://doi.org/10.1088/13616579/aab6d0

7. Barcelo, C.H., Alvarez, J.M., Shakir, M.Z., Alcaraz-Calero, J.M., Casasecade-la-Higuera, P.: Efficient k-NN implementation of real-time detection of cough events in smartphones. IEEE J. Biomed. Health Inf. **22**(5), 1662–1671 (2018). https://doi.org/10.1109/jbhi.2017.2768162

8. Monge-Alvarez, J., Barcelo, C.H., San-Jose-Revuelta, L.M., Casaseca-de-laHiguera, P.: A machine hearing system for robust cough detection based on a highlevel representation of band-specific audio features. IEEE Trans. Biomed. Eng. **66**(8), 2319–2330 (2019). https://doi.org/10.1109/tbme.2018.2888998

9. Sun, X., Lu, Z., Hu, W., Cao, G.: SymDetector: detecting sound-related respiratory symptoms using smartphones. In: UbiComp 2015: Proceedings of the 2015 ACM International Joint Conference on Pervasive and Ubiquitous Computing, pp. 97–108, September 2015. https://doi.org/10.1145/2750858.2805826

10. Amoh, J., Odame, K.: Deep neural networks for identifying cough sounds. IEEE Trans. Biomed. Circ. Syst. **10**(5), 1003–1011 (2016). https://doi.org/10.1109/tbcas.2016.2598794

11. Kadambi, P., et al.: Towards a wearable cough detector based on neural networks. In: 2018 IEEE International Conference on Acoustics, Speech and Signal Processing (ICASSP), April 2018. https://doi.org/10.1109/icassp.2018.8461394

12. Amoh, J., Odame, K.: DeepCough: a deep convolutional neural network in a wearable cough detection system. In: 2015 IEEE Biomedical Circuits and Systems Conference (BioCAS), Atlanta, GA, pp. 1–4, September 2015. arXiv:1509.02512

13. Tirzite, M., Bukovskis, M., Strazda, G., Jurka, N., Taivans, I.: Detection of lung cancer in exhaled breath with an electronic nose using support vector machine analysis. J. Breath Res. **11**(3), 036009 (2017). https://doi.org/10.1088/17527163/aa7799

14. Yahya, O., Faezipour, M.: Automatic detection and classification of acoustic breathing cycles. In: Proceedings of the 2014 Zone 1 Conference of the American Society for Engineering Education, Bridgeport, CT, pp. 1–5, May 2014. https://doi.org/10.1109/aseezone1.2014.6820648

15. Larson, E.C., Goel, M., Boriello, G., Heltshe, S., Rosenfeld, M., Patel, S.N.: SpiroSmart: using a microphone to measure lung function on a mobile phone. In: UbiComp 2012, pp. 280–289, September 2012. https://doi.org/10.1145/2370216.2370261

16. Windmon, A., Minakshi, M., Chellappan, S., Athilingam, P.R., Johansson, M., Jenkins, B.A.: On detecting COPD cough using audio signals recorded from smart-phones. In: Proceedings of 11th International Joint Conference on Biomedical Engineering Systems and Technologies: HEALTHINF, vol. 5, pp. 329–338, February 2018. https://doi.org/10.5220/0006549603290338

17. Windmon, A., et al.: TussisWatch: a smart-phone system to identify cough episodes as early symptoms of chronic obstructive pulmonary disease and congestive heart failure. IEEE J. Biomed. Health Inf. **23**(4), 1566–1573 (2019). https://doi.org/10.1109/jbhi.2018.2872038

18. Ramalho, G.L.B., Filho, P.P.R., Sombra de Medeiros, F.N., Cortez, P.C.: Lungdisease detection using feature extraction and extreme learning machine. Rev. Bras. Eng. Biomed. 30(3), 207–214 (2014). https://doi.org/10.1590/rbeb.2014.019

19. Waltisberg, D., Amft, O., Brunner, D.P., Troster, G.: Detecting disordered breathing and limb movement using in-bed force sensors. IEEE J. Biomed. Health Inf. **21**(4), 930–938 (2017). https://doi.org/10.1109/jbhi.2016.2549938

20. Korpas, J., Sadlonova, J., Vrabec, M.: Analysis of cough sound: an overview. Pulm. Pharmacol. **9**, 261–268 (1996). https://doi.org/10.1006/pulp.1996.0034
21. Chatrzarrin, H., Arcelus, A., Goubran, R.: Feature extraction for the differentiation of dry and wet cough sounds. In: Proceedings of IEEE International Symposium on Medical Measurements and Applications, pp. 162–166, July 2011. https://doi.org/10.1109/memea.2011.5966670
22. Slok, A.H.M., et al.: Development of the assessment of burden of COPD tool: an integrated tool to measure the burden of COPD. NPJ Prim. Care Respir. Med. **24**(1–4), 14021 (2014). https://doi.org/10.1038/npjpcrm.2014.21
23. Birring, S.S., Spinou, A.: How best to measure cough clinically. Curr. Opin. Pharmacol. **22**, 37–40 (2015). https://doi.org/10.1016/j.coph.2015.03.003
24. Muda, L., Begam M., Elamvazuthi, I.: Voice recognition algorithms using melfrequency cepstral coefficient (MFCC) and dynamic time warping (DTW) techniques. J Comput. **2**(3), 138–143 (2010). arXiv:1003.4083
25. Logan, B.: Mel frequency cepstral coefficients for music modeling. In: International Symposium on Music Information Retrieval (ISMIR) (2000). http://citeseerx.ist.psu.edu/viewdoc/summary?doi=10.1.1.11.9216
26. Lyons, J.: Mel frequency cepstral coefficient (MFCC) tutorial (2012). http://practicalcryptography.com/miscellaneous/machinelearning/guide-mel-frequency-cepstral-coefficients-mfccs/
27. Nisar, S., Shahzad, I., Khan, M.A.: Pashto spoken digits recognition using spectral and prosodic based feature extraction. In: 2017 Ninth International Conference on Advanced Computational Intelligence (ICACI), Doha, pp. 64–78, February 2017. https://doi.org/10.1109/icaci.2017.7974488
28. Cortes, C., Vapnik, V.: Support-vector networks. J. Mach. Learn. **20**(3), 273–297 (1995). https://doi.org/10.1023/a:1022627411411
29. Leidy, N.K., et al.: Development of the exacerbations of chronic obstructive pulmonary disease tool (EXACT): a patient-reported outcome (PRO) measure. Value Health: J. Int. Soc. Pharmacoecon. Outcomes Res. **13**(8), 965–975 (2010). https://doi.org/10.1111/j.1524-4733.2010.00772.x
30. Leidy, N.K., Murray, L.T.: Patient-reported outcome (PRO) measures for clinical trials of COPD: the EXACT and E-RS. COPD: J. Chron. Obstruct. Pulm. Dis. **10**(3), 393–398 (2013). https://doi.org/10.3109/15412555.2013.795423
31. Wac, K., Hausheer, D.: COPD24: from future internet technologies to health telemonitoring and teletreatment applications. In: 12th IFIP/IEEE International Symposium on Integrated Network Management and Workshops, Dublin, pp. 812–826, August 2011. 978-1-4244-9221-3
32. Ding, H., Moodley, Y., Kanagasingam, Y., Karunanithi, M.: A mobile-health system to manage chronic obstructive pulmonary disease patients at home. In: 34th Annual International Conference of the IEEE Engineering in Medicine and Biology Society, San Diego, CA, pp. 2178–81, September 2012. https://doi.org/10.1109/embc.2012.6346393
33. Smith, J.A., Ashurst, H.L., Jack, S., Woodcock, A.A., Earis, J.E.: The description of cough sounds by healthcare professionals. Cough **2**(1) (2006). https://doi.org/10.1186/1745-9974-2-1
34. Self, D.: Audio Engineering Explained, 1st edn, pp. 200 & 446. Focal Press, Burlington (2012). ISBN 0240812735. E-Book
35. Huang, X., Acero, A., Hon, H.: Spoken Language Processing: A Guide to Theory, Algorithm, and System Development, 1st edn. Prentice Hall, Upper Saddle River (2001). ISBN-10: 0130226165
36. DeMeo, D.L., et al.: Women manifest more severe COPD symptoms across the life course. Int. J. Chron. Obstruct. Pulm. Dis. **13**, 3021–3029 (2018). https://doi.org/10.2147/copd.s160270

37. Xu, W., et al.: mCOPD: mobile phone based lung function diagnosis and exercise system for COPD. In: ACM International Conference on Pervasive Technologies Related to Assistive Environments (PETRA 2013) Proceeding Series, Rhodes, Greece, May 2013. https://doi.org/10.1145/2504335.2504383

38. Kopparapu, S.K., Laxminarayana, M.: Choice of mel filter bank in computing MFCC of a resampled speech. In: 10th International Conference on Information Sciences, Signal Processing and their Applications (ISSPA, 2010), Kaula Lumpur, pp. 121–124, May 2010. https://doi.org/10.1109/isspa.2010.5605491

39. French, P., Foulkes, P., Harrison, P., Hughes, V., Segundo, E.S., Stevens, L.: The vocal tract as a biometric: output measures, interrelationships, and efficacy. In: Proceedings of ICPhS, University of Glasgow, vol. 18 (2015). ISBN 978-0-85261-941-4

40. Serizel, R., Giuliani, D.: Vocal tract length normalisation approaches to DNN based children's and adults' speech recognition. In: 2014 IEEE Spoken Language Technology Workshop (SLT), South Lake Tahoe, NV, pp. 135–140, April 2015. https://doi.org/10.1109/slt.2014.7078563

41. Singh, N., Khan, R.A., Shree, R.: MFCC and prosodic feature extraction techniques: a comparative study. Int. J. Comput. Appl. 54(1), 9–13 (2012). https://doi.org/10.5120/8529-2061

42. Cassidy, R.J.: Audio Speech Research Note. Stanford University, July 2003. https://ccrma.stanford.edu/rjc/pubs/audio-speech/audio-speech.pdf

43. Terasawa, H., Berger, J., Makino, S.: In search of a perceptual metric for timbre: dissimilarity judgments among synthetic sounds with MFCC-derived spectral envelopes. J. Audio Eng. Soc. 60(9). 674–685 (2012). http://www.aes.org/elib/browse.cfm?elib=16372

44. Rabiner, L., Juang, B.H.: Fundamentals of Speech Recognition, 1st edn, pp. 183–190. Pearson, London (1993). ISBN-13: 978-0130151575

45. Liu, J.M., et al.: Cough signal recognition with gammatone cepstral coefficients. In: 2013 IEEE China Summit and International Conference on Signal and Information Processing, Beijing, China, pp. 160–164, July 2013. https://doi.org/10.1109/chinasip.2013.6625319

46. Pramono, R.X.A., Imtiaz, S.A., Rodriguez-Villegas, E.: Automatic cough detection in acoustic signal using spectral features. In: 2019 41st Annual International Conference of the IEEE Engineering in Medicine and Biology Society (EMBC), Berlin, Germany, October 2019. https://doi.org/10.1109/embc.2019.8857792

47. Abushakra, A., Faezipour, M.: Acoustic signal classification of breathing movements to virtually aid breath regulation. IEEE J. Biomed. Health Inform. 17(2), 493–500 (2013). https://doi.org/10.1109/jbhi.2013.2244901

48. Palaniappan, R., Sundaraj, K., Lam, C.K.: Reliable system for respiratory pathology classification from breath sound signals. In: 2016 International Conference on System Reliability and Science, Paris, pp. 152–156, January 2017. https://doi.org/10.1109/icsrs.2016.7815855

49. Lin, B.-S., Lin, B.-S.: Automatic wheezing detection using speech recognition technique. J. Med. Biol. Eng. 36(4), 545–554 (2016). https://doi.org/10.1007/s40846-016-0161-9

50. Smith, J.A., Ashurst, H.L., Jack, S., Woodcock, A.A., Earis, J.E.: The description of cough sounds by healthcare professionals. Cough 2(1), 1–9 (2006). https://doi.org/10.1186/1745-9974-2-1

51. Jensen, J.H., Christensen, M.G., Murthi, M.N., Jensen, S.H.: Evaluation of MFCC estimation techniques for music similarity. In: 14th European Signal Processing Conference, Florence, Italy, September 2006. http://www2.imm.dtu.dk/pubdb/edoc/imm4413.pdf

52. Mohan, B.J., Babu, R.N.: Speech recognition using MFCC. In: International Conference on Advances in Electrical Engineering (ICAEE), Vellore, India, pp. 1–4, January 2014. https://doi.org/10.1109/icaee.2014.6838564

53. Shi, Y., Liu, H., Wang, Y., Cai, M., Xu, W.: Theory and application of audio based assessment of cough. Hindawi J. Sens. **2018**, 1–8 (2018). Article ID 9845321. https://doi.org/10.1155/2018/9845321
54. Boujelben, O., Bahoura, M.: Efficient FPGA-based architecture of an automatic wheeze detector using a combination of MFCC and SVM algorithms. J. Syst. Arch. **88**, 54–64 (2018). https://doi.org/10.1016/j.sysarc.2018.05.010
55. Audacity Manual: Sample Rates, November 2019. https://manual.audacityteam.org/man/sample-rates.html

Safe Navigation by Vibrations on a Context-Aware and Location-Based Smartphone and Bracelet Using IoT

Rusho Yonit[✉], Elbaz Haim, Leib Reut, and Polisanov Roni

Software Engineering Department, Shenkar College, Anne Frank St 12, Ramat Gan, Israel
yonit.rusho.17@gmail.com

Abstract. How can the use of IoT improve safely navigation for kick scooters? The context-aware and location-based solution combines an embedded smart bracelet and a mobile phone. The integrated platform provides a ubiquitous service, which gives kick scooters the ability to safely navigate by reducing distractions.

A key motivation was to minimize attention to the external disturbances during navigation and maximize attention to the route itself. This was achieved by (1) providing guidance while navigating via synchronized vibro-tactile feedback on a smart bracelet; (2) developing a vibration language on the bracelet; (3) allowing users to produce navigation routes in-advanced on their mobile phone, or consume existing routes in real-time.

The solution was designed for inclusive and tested in real-life situations on busy roads in the city.

Keywords: Ubiquitous and mobile computing · Location based services · Contextual design · Accessibility

1 Introduction

Nowadays, kick scooters have become an integral part of teenagers' daily lives. Research shows that a massive part of road accidents involve two-wheeled vehicle [10]. Kick scooter accidents in school-aged children and adolescents are associated with road accidents [17]. Indeed, the number of patients visiting emergency department due to push scooter accidents increases [9].

Regarding 4-wheels transportation, studies associate mobile phone use to road accidents. Driving while using the mobile phone reduces safety due to cognitive load [6]. Researches proven connection between mobile phone use while driving and traffic safely [8]. On the other hand, drivers who seek for navigation direction need to divide their attention between the road and the mobile device for the purpose of getting directions. Even though distracted drivers are involved in varies levels of self-regulation [11], psychological factors and attitude still determine some mobile phone use while driving [18]. This unsafe behavior is also typical of pedestrians, cyclists and electric or mechanical scooters.

J. Liu et al. (Eds.): MobiCASE 2020, LNICST 341, pp. 121–133, 2020.
https://doi.org/10.1007/978-3-030-64214-3_8

Therefore, the current study places kick scooter safety at the limelight and explores the effects of Internet of Things (IoT) solutions on safe navigation. First, in order to reduce the use of mobile phone for navigation while riding a kick scooter, a wearable computing will be used. Second, the solution should be designed as a location-based service (LBS) which integrates geographic location (as coordinates) with applications, such as emergency services, car navigation systems or tourism tour planning. These applications run on computers, personal assistants, phones and provide users with added value to mere location information [16]. Third, due to the combination of ubiquitous computing via mobile phone and wearable devices, the use of context is increasingly important where the users' context is changing rapidly [1]. Such changes are relevant in the context of driving and navigating. Fourth, Inclusive Design helps creating products that serve as many people as possible and enables people with diverse characteristics to use a product in a variety of different environments [13].

Therefore, the influence of IoT is examined while implementing four principles: (1) A wearable smart bracelet; (2) A location-based technique, (3) A context-awareness perspective implemented on mobile phone device; (4) Inclusive design approach.

Following the four guidelines, this paper presents the processes of design and implementation of a comprehensive prototype that integrates the four fundamentals: wearable, location-based, context-awareness mobile interface and inclusive design. The solution includes a wearable bracelet for vibrotactile feedback in a guidance systems. Hence, a location-based mobile phone sends GPS data and in-advanced planned route information through the Internet to an embedded smart bracelet. The bracelet transmits the data to navigation guidance and vibrates according to a vibration language, described later. The system allows users to plan the route in-advanced by selecting an existing route or creating a new customized one quickly. All routes are saved in the cloud for future use and customization. Based on saved and analyzed data, the system calculates and offers routes which are optimized to users' preferences

Finally, the system supports inclusive design. People with disabilities can navigate their way from starting point to destination wearing the bracelet and follow vibration guidance while navigating their way.

The next section describes related work.

2   Related Work

Technical solutions for optimized, safe and efficient navigation are evolving. One of the goals of ubiquitous computing, is to make devices 'smart' [2]. One example is a vehicle navigation solution for urban environment using IoT [5]. A second example is XIAOMI BAND, which is a wristband that connects to a mobile device and offers navigation functionality to the user. The XIAOMI is based on cutting-edge technologies and advanced hardware. It can convert vibration alert and digital display. Another example is the SUNU, which is a bracelet that offers a navigation system for the visually impaired or hearing impaired by a combination solution of the bracelet and laminated reality glasses. A recent study explores a vibro-tactile feedback around user's wrist to convey information about a general direction of a target. Instead of defining a route, this study enables free exploration of the surrounding [4]. The current study explicitly highlights safety while navigating to a specific target. Following the above examples of wearable solutions and smart navigation systems, the first two non-functional requirements are:

The Solution Aims to be Designed as a Smart Wearable Product. Meaning, a complementary wearable device, which is activated using a combined application.

The Solution Aims to be Designed as a Location-Based Mobile Phone.
Meaning, using GPS, the solution integrates geographic location. Information will be shown to the user based on local geographical location on his or her mobile device.

Research shows that vibrotactile wristband delivering messages and information is efficient in order to reduce accidents [3]. In case of warning messages, the vibrations assist in reducing street-crossing risks for all users. Placing user in the center has been shown to be an effective method to ensure that the systems being developed meets users' expectations [12]. Following are the next two non-functional requirements:

The Solution Aims to be Designed as a Context-Awareness Solution. Meaning, the mobile application and bracelet should be aware of the dynamic environment and change accordingly with minimum delays. Notifications, instructions and other information should be displayed or vibrated on-demand.

The Solution Aims to be Designed for Inclusion. Meaning, the product should serve as many people as possible, including teenagers and older, people with hearing disabilities or with cognitive difficulties, etc.

Having explained the non-functional requirements of the solution, the following section elaborates on the content and information being produced and consumed using the product.

2.1 Information Consumption Process

The information being used before and during navigation may range from complex visualization maps on the mobile phone, to simple notification alerts on the bracelet. Information is a peculiar product in that anyone can act as consumer and producer. Each role in the information cycle may lead to differing user-experiences. Therefore, user-experience is treated in this research in two perspectives. The first is the consumer's perspective- consumer's interface design is composed, among others, of content. The content is what the interface presents to users, including text and media [7]. According to this approach, consumers use existing information. In the current research they can search, select and navigate in real-time between existing tracks.

2.2 Information Production Process

The second perspective is from a producer's point of view: the user-experience of producing information via the mobile phone interface. Producers in the current research can create information, such as routes in-advanced, aggregate existing routes or customize tracks. The process of information production is a five-steps process: specification, design, implementation, validation and evolution [14, 15]. The specification step includes determine target audience (impatient youth in grade school, or a hearing impaired person, etc.). In addition, this first step includes understanding the goal and constrains of the produced information. For example, to navigate from home to school safely. The design step includes planning how the consumers will use the produced information. For example, the track should be customized for inclusion. The development step is the actual creation of the information. Meaning, the definition starting point and destination, and choosing the safer path. The validation ensures that produced information fulfills consumers' expectations and needs and that mistakes are minimal. For example, if the goal is to maximize safety for children riding on their kick scooter to school, then through the validation step, the producer should go backwards and verify that the produced track is the safest path for the child, before the child actually uses the track in real-time.

The final step is evaluating the produced information. It can be performed by the producer testing it in real-time on the road in busy streets, to evaluate its safety level.

Having explained the roles of information consumption and production in the context of a location-based, context-awareness solution and design for inclusion, the main functional requirement is as following:

The solution should allow the user to produce information in advanced and consume it in real-time.

The name of the integrated solution is LookUp. The following sections elaborate on its design and implementation.

3   Solution Approach

With the goal to maximize safety on the road, the system approach is to eliminate the need to lower one's eyes out of the way to look at the mobile phone while navigating. LookUp solution embraces IoT methods for communication between smart devices, such as a mobile phone and a smart bracelet. In addition, the system adapts context-awareness definitions, such as automatic interface and offers changes according to both personal user preferences and synchronized notifications based on location and environment.

The solution offers safe navigation for two types of users: track-consumers and track-producers. Track consumer uses the system to orient and navigate using maps displayed with the smart bracelet. These real-time users can wear the bracelet component on the hand and feel the direction indication with real-time vibrations.

Track producers create routes in-advanced on their mobile phone. The application allows customizing an existing route, or alternatively producing a new route from starting point to destination. Producers can search for existing tracks in the repository for aggregation purpose, produce an automatic generator path and control the produced route. Moreover, producers can automatically view the map according to current location and evaluate the produced information.

Following the above, it can be understood that the system is divided to two processes, one for information consumer and the other for information producer. Hence, the system architecture is divided into two phases of the life cycle: predefined activities and real-time activities. Figure 1 displays system overview, focusing on predefined activities. Figure 2 focuses on real-time navigation during activity.

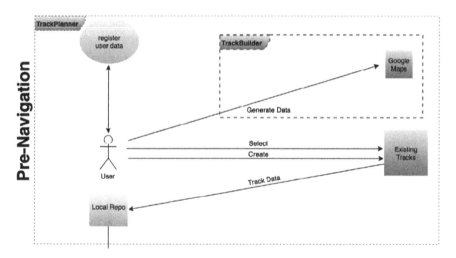

Fig. 1. System overview: predefined activities

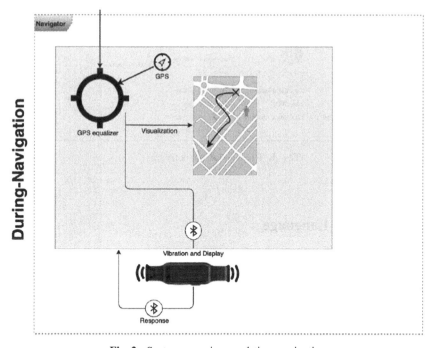

Fig. 2. System overview: real-time navigation

Figure 3 describes roles, data and activities in the integrated solution, separated to pre navigation and during navigation processes. In addition, Fig. 3 displays the various activities for the main roles: track producer and track consumer. The platform uses Maps

API'S combined with Bluetooth technology, GPS location, and the available activities for users when in-advanced planning and during navigating a route.

Fig. 3. Roles, data and activities

4   The Vibration Language

The study involved two steps before developing the vibrational language. The first phase included a preliminary pilot to assess the feasibility and efficiency of the language. We conducted the second phase after the development of a prototype. The purpose of the second experiment was to examine the solution before full development.

4.1   Participants and Experiments

The first pilot involved 25 participants, randomly selected on the street. We chose a partially crowded street in a central city in the country. Participants were of ages 20 to 40 and were informed in-advanced about the purpose of the study. Following the preliminary pilot, we came to the realization that a vibration language using signals on one side of the bracelet is not attractive nor effective.

Thus, in the second stage, we enlarged routes indications by signaling on two sides of the bracelets and by adding a small screen with additional information. We developed a friendly and easy to remember language for the navigation signals.

After an initial development and with the first prototype of the product, we approached 15 random people on the street. We invited them to use the product and experience the live navigation. Following their responses, we concluded results: 12 people out of 15 expressed interest in using the bracelet for safe navigation (three people expressed no interest).

4.2 The Language

Table 1 describes the vibration language. The number of vibrations reflects the distance-as closer the rider gets to the turn, the more vibrations accrue. For example, twenty meters before a turn, the bracelet vibrates once. Five meters before the turn, the bracelet vibrates twice. When the rider reaches the turn, the bracelet vibrate three times. This behavior is similar to the phone ringing behavior. At first it is weak and over time the sound becomes stronger. Alarm clock also starts with a weak sound and over time increases. When driving back in the car, the beep sound of the alarm device increases as the car approaches the object behind the vehicle. In translation to the vibrations language, as the turn approaches, the vibrations increase.

Table 1. Vibration language.

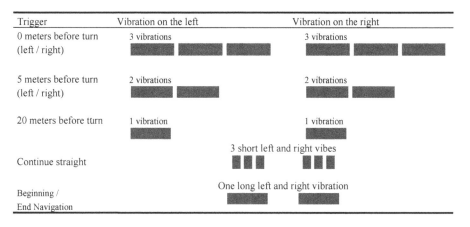

Trigger	Vibration on the left	Vibration on the right
0 meters before turn (left / right)	3 vibrations	3 vibrations
5 meters before turn (left / right)	2 vibrations	2 vibrations
20 meters before turn	1 vibration	1 vibration
Continue straight	3 short left and right vibes	
Beginning / End Navigation	One long left and right vibration	

5 System Overview

This section details the development environment and programming languages. On the client side, the system is developed using JavaScript (JS) React library, which uses Virtual DOM for faster updates. Moreover, React supports data binding technique that enables synchronization between the properties of two separated objects and thus facilitates the transfer of information to the bracelet in real time. In addition, LookUp uses external APIs such as Google Maps API and Google Directions API.

On the server side, the system is developed using Node.js, which interprets JS written in C++ and based on a V8 engine, runs on a Chrome browser. LookUp implements modern RESTful architecture on the server side, which is suited to the needs of the system and server side technology.

The bracelet is based on Arduino C using built-in i2c display libraries. The Bluetooth component through which the communication is made is HC-08, which supports version 4.0 to match the support with an Internet browser. The interface is performed by setting up a serial port to connect, receive and process data received therein.

5.1 In-Advanced Planning Routes on Mobile Phones

Figure 4 displays planning in advanced route in a mobile phone device. In specific, it displays Custom generate track page in the planning process. The interface is responsive for tables and desktop devices.

Fig. 4. Planning route in-advanced

Once the track producer generates a route, the system is ready for the track consumer for real-time navigation.

The next section details activities available during navigation.

5.2 The Bracelet for Real-Time Navigation

This section describes the embedded part of the integrated solution. The bracelet is aware of user location and context, therefore it vibrates based on route and dynamic environment.

LookUP was first developed on an Arduino UNO. After fully developing its software and prototype hardware, the goal was to minimize the solution. Figure 5 displays the initial prototype, which was not suitable to be a wearable device due to its large size. The second phase included adjustments and performance evaluations in order to create a wearable fully working component. Eventually, the final product's minimized and optimized schema is presented in Fig. 6. It is based on an Arduino pro mini microprocessor, Bluetooth component HC-08 in order to support BLE v4.0 (which is necessary for Web BLE support). Equipped with an OLED 0.96 Inch screen for visual indications, two vibration motors and a 3.7 V lithium battery which provides eight hours of continuous work.

For the purpose of inclusive design support, the bracelet fits a variety of users with different body figures. Its tough plastic casing emphasis the vibrations in order for the user to identify the vibrations language. The bracelet's rubber customizable bands are adjustable and exchangeable in order to provide the user with the comfort needed while navigating. Figure 7 displays the smart bracelet, in its final look. Figure 8 is a live camera frame, taken when tested the final version of the system. A teenager is riding the kick scooter wearing the smart bracelet and navigating in real time on the road.

Fig. 5. The initial prototype

The next section elaborates on the synchronization between the mobile and bracelet, and their connectivity to the cloud.

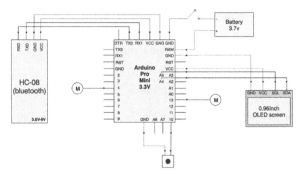

Fig. 6. Smart bracelet - electrical scheme

Fig. 7. The final smart bracelet

Fig. 8. Real time alerts

5.3 Synchronization Between Mobile, Cloud and Bracelet

Synchronization between the mobile phone, the cloud for saving and analyzing data; and the smart bracelet is critical for real-time data and navigation decision making.

The interactions between core modules and components includes: Track Planner module, Track Builder module and Navigation controller. Track Planner handles interaction with the user, synchronizes user preferences with Google Maps API and handles data retrieval from the database. The Navigation controller manages the navigation process. It communicates requests from Track Planner and forwards requests to Track Builder. In addition, the Navigation controller sends route details and directions to the embedded bracelet. The Track Builder interacts with Google Directions API and creates new tracks requested by Track Planner (Fig. 9).

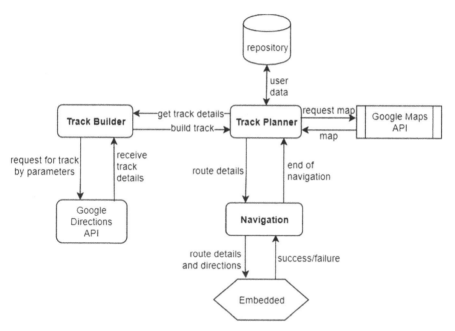

Fig. 9. Interaction between core modules

In summary, the system is divided into several key parts, each of which plays a significance role in the LookUp system, enabling reducing network traffic loads to maintain real-time information transfer.

6 Limitations

In order to create a custom itinerary, it is necessary to design a route as a separated layer on the map. In addition, access to the Internet is required in order to send requests and receive responses from the cloud and the mobile phone. In addition, it is required to keep continuous Bluetooth connectivity from the mobile device to a component throughout the navigation path. A non-functional requirement is the need to immediately generate vibrations and keep synchronization with the navigation path. Therefore, minimum latency as possible is required. In order to activate the system, the user must have a mobile device that allows the use of location data using a built-in GPS antenna. Moreover, the user must have a mobile device that supports Bluetooth version 4.2 or higher.

7 Discussion and Conclusion

With the aim of finding if the use of IoT improves safely navigation for kick scooters, this study defined major nonfunctional and functional requirements, validated the process of design and implementation and eventually presents a context-aware and location-based solution combines an embedded smart bracelet and a mobile phone. Moreover, the comprehensive product provides a ubiquitous application, which gives kick scooters the ability to safely navigate by reducing distractions.

The main nonfunctional requirements were fully implemented and evaluated: (1) the solution was designed as a smart wearable product; (2) it was designed as a smart wearable product; (3) the solution was designed as a context-awareness application; (4) The solution was designed for inclusion. In addition, the core functional requirement was to allow the user to produce information in advanced and consume it in real-time.

Attention to external disturbances was minimized during navigation. In addition, attention was maximized to the consumed route. This was achieved by (1) providing guidance while navigating by vibration feedback on a smart bracelet; (2) creating a vibration language and evaluated it in real-time; (3) allowing users to produce and share navigation routes in advanced on their mobile phone, or consume existing routes in real time.

Aiming to offer a safe navigation alternative, while keeping inclusive design rules, required examination of two key issues during the process. The first issue is the vibration language. For this purpose, we conducted two experiments. The conclusion from the analysis of the results is that signaling only on one side does not allow defining a rich language for real-time navigation. The second issue refers to the technical alternatives. After examining related location-based solutions, and examining context-awareness applications, the conclusion was to combine the two methods: location-based and context-awareness, into an integrative solution using a smart bracelet with a smartphone. Data transfers between the two devices, stored in the cloud, and analyzes for learning and improvement. In addition, enhancing the application with options to pre produce routes, save favorites routes, enable both automatic and custom creation of routes, enriches the solution for the consumers in real-time situation.

References

1. Abowd, G.D., Dey, A.K., Brown, P.J., Davies, N., Smith, M., Steggles, P.: Towards a better understanding of context and context-awareness. In: Gellersen, H.-W. (ed.) HUC 1999. LNCS, vol. 1707, pp. 304–307. Springer, Heidelberg (1999). https://doi.org/10.1007/3-540-48157-5_29
2. Basta, N., El Nahas, A., Grossmann, H.-P., Abdennadher, S.: Guess where I go? In: Proceedings of the 17th International Conference on Mobile and Ubiquitous Multimedia - MUM 2018, pp. 93–102 (2018). https://doi.org/10.1145/3282894.3282911

3. Cœugnet, S., et al.: A vibrotactile wristband to help older pedestrians make safer street-crossing decisions. Accid. Anal. Prev. **109**(December 2017), 1–9 (2017). https://doi.org/10.1016/j.aap.2017.09.024

4. Dobbelstein, D., Rukzio, E., Henzler, P.: Unconstrained pedestrian navigation based on vibrotactile feedback around the wristband of a smartwatch. In: Conference on Human Factors in Computing Systems - Proceedings, pp. 2439–2445 (2016). https://doi.org/10.1145/2851581.2892292

5. Godavarthi, B., Nalajala, P., Ganapuram, V.: Design and implementation of vehicle navigation system in urban environments using internet of things (Iot). In: IOP Conference Series: Materials Science and Engineering (2017). https://doi.org/10.1088/1757-899X/225/1/012262

6. Lamble, D., Kauranen, T., Laakso, M., Summala, H.: Cognitive load and detection thresholds in car following situations: Safety implications for using mobile (cellular) telephones while driving. Accid. Anal. Prev. **31**(6), 617–623 (1999). https://doi.org/10.1016/S0001-4575(99)00018-4

7. Lee, Y.E., Benbasat, I.: Interface design for mobile commerce. Commun. ACM **46**(12), 48 (2003). https://doi.org/10.1145/953460.953487

8. Lipovac, K., Ðerić, M., Tešić, M., Andrić, Z., Marić, B.: Mobile phone use while driving-literary review. Transp. Res. Part F Traffic Psychol. Behav. **47**, 132–142 (2017). https://doi.org/10.1016/j.trf.2017.04.015

9. Mebert, R.V., Klukowska-Roetzler, J., Ziegenhorn, S., Exadaktylos, A.K.: Push scooter-related injuries in adults: an underestimated threat? Two decades analysed by an emergency department in the capital of Switzerland. BMJ Open Sport Exerc. Med. **4**(1) (2018). https://doi.org/10.1136/bmjsem-2018-000428

10. Nitin, C.M., Patil, S.: A retrofit design of safety and stability mechanism for two wheelers. In: IOP Conference Series: Materials Science and Engineering (2019). https://doi.org/10.1088/1757-899X/594/1/012035

11. Oviedo-Trespalacios, O., Haque, Md.M., King, M., Washington, S.: Mate! I'm running 10 min late": an investigation into the self-regulation of mobile phone tasks while driving. Accid. Anal. Prev. **122**(January), 134–142 (2019). https://doi.org/10.1016/j.aap.2018.09.020

12. Perebner, M., Huang, H., Gartner, G.: Applying user-centred design for smartwatch-based pedestrian navigation system. J. Locat. Based Serv. **13**(3), 213–237 (2019). https://doi.org/10.1080/17489725.2019.1610582

13. Persson, H., Åhman, H., Yngling, A.A., Gulliksen, J.: Universal design, inclusive design, accessible design, design for all: different concepts—one goal? On the concept of accessibility—historical, methodological and philosophical aspects. Univ. Access Inf. Soc. **14**(4), 505–526 (2014). https://doi.org/10.1007/s10209-014-0358-z

14. Rusho, Y., Raban, D.R.: Hands on: information experiences as sources of value. J. Assoc. Inf. Sci. Technol. (2019). https://doi.org/10.1002/asi.24288

15. Rusho, Y., Raban, D.R.: The effects of information production process on experience and evaluations. In: The 19th Annual Conference of the Association of Internet Researchers (AoIR 2018: Transnational Materialities Montreal, Canada, 10–13 October 2018) (2018)

16. Schiller, J., Voisard, A.: Location-Based Services. Elsevier, Amsterdam (2004)

17. Unkuri, J.H., Salminen, P., Kallio, P., Kosola, S.: Kick scooter injuries in children and adolescents: minor fractures and bruise. Scand. J. Surg. **107**(4), 350–355 (2018). https://doi.org/10.1177/1457496918766693

18. Walsh, S.P., White, K.M., Hyde, M.K., Watson, B.: Dialling and driving: factors influencing intentions to use a mobile phone while driving. Accid. Anal. Prev. **40**(6), 1893–1900 (2008). https://doi.org/10.1016/j.aap.2008.07.005

Key Location Discovery of Underground Personnel Trajectory Based on Edge Computing

Zhao Jinjin[1,2,3], Zou Xiangyu[1,2,3(✉)], Zhang Yu[1,2,3], Gu Youya[1,2,3], Wu Fan[4], and Zhu Zongwei[4]

[1] School of Information and Control Engineering, China University of Mining and Technology, Xuzhou 221008, Jiangsu, China
hainuoeileen@163.com

[2] Internet of Things (Perception Mine) Research Center, China University of Mining and Technology, Xuzhou 221008, Jiangsu, China

[3] The National Joint Engineering Laboratory of Internet Applied Technology of Mines, Xuzhou 221008, Jiangsu, China

[4] Suzhou Institute for Advanced Study, University of Science and Technology of China, Suzhou 215000, China

Abstract. With the rapid development of smart mines, workers' trajectories can be accurately tracked and generate massive positioning data. However, how to quickly find useful information in a large amount of data is an important issue at present. Consequently, in this paper, we propose an algorithm for application in underground edge computing systems, called key location discovery (KLD). First, the algorithm reconstructs the trajectory data by the potential semantic information of the underground environment and miners work types to be more suitable for the actual situation. Then, the KLD algorithm screen out the key locations of underground personnel trajectories according to inflection point and stay point. In the meanwhile, compared with the trajectory structure-based hot spots (TS_HS) discovery algorithm, KLD algorithm reduced the positioning data by 1/4 and calculating time. In addition, placing the algorithm proposed in this paper on the edge side for calculation and processing can filter out key information in real time, which is more beneficial to the follow-up work, including the study of personnel trajectory abnormalities and prediction.

Keywords: Personnel trajectories analysis · Inflection point · Stay point · Edge computing

1 Introduction

With the rapid development of the mining economy, people have higher and higher requirements for the safety of underground traffic and the speed of data

Supported by the National Key Research and Development Program of China (2017YFC 0804402).

J. Liu et al. (Eds.): MobiCASE 2020, LNICST 341, pp. 134–142, 2020.
https://doi.org/10.1007/978-3-030-64214-3_9

processing. Currently, various positioning equipments such as radio frequency identification, ZigBee [2], WiFi [6], Bluetooth [5], and UWB [3] have been placed in the coal mines and its surrounding areas. Therefore, an increasing quantity of positioning data can be gathered and transmitted by positioning stations underground to the ground server. However, the method of transferring data to the server takes a lot of time, causing delay in information processing, so that the movement and safety status of underground personnel cannot be observed in real time.

Therefore, combining the advantages of edge computing, this paper proposes an algorithm for data processing on the edge side, that is, key location discovery, which can quickly filter out useful information and find key positions. First, considering the semantic information of underground workers, the paper classifies the trajectories of the same type of work according to the miners' job information, which is more suitable for the actual situation. Then, the KLD algorithm reconstructs the trajectory data, and it can screen out the key locations of underground personnel trajectories according to inflection point and stay point. Compared with other algorithms, The KLD algorithm reduced the positioning data by 1/4, thereby removing redundant data.

2 Materials and Methods

A key position underground is a point with special semantics in the trajectory, such as crossroads and work areas. These positions are points with high trajectory density. However, a high-density point is not necessarily a key position point. The trajectory of the wellhead is relatively dense, but the wellhead does not have much practical significance for the overall trajectory of the moving object. Therefore, the emphasis of this chapter is to filter the key positions sequences in the massive trajectory data, that is, a sequence of positions or regions with special semantics in the trajectory. For actual situations in the well, the KLD algorithm flowchart is shown in Fig. 1. First, because different types of work have different activity areas and characteristics, the paper divides the discrete positioning points into different personnel trajectories with semantic information according to the type of work information, which is convenient for subsequent work. Then, different types of trajectory points are used to determine whether it is an inflection point according to the turning angle. If the result is positive, it is added to the key sequence. Otherwise, it continues to be used to judge whether it is a stay point until all the trajectory points are traversed. The important semantic points in the trajectory are extracted to form a key position sequence, thereby reducing data redundancy and calculation complexity.

2.1 Personnel Semantics

Because of the different divisions of labour, the underground workers' trajectories have their own characteristics. Table 1 shows part of the positioning data log. The trajectory data consist of a series of orbital positioning points, including

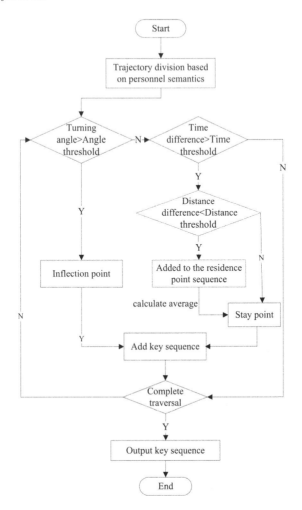

Fig. 1. KLD algorithm flowchart.

Table 1. Positioning data log

Number	Type of work	Name	Initial reporting time	Last reporting time	Location number
1	Group I comprehensive Coal Mining Team	Mr. Chen	2017-12-20 06:05:42	2017-10-2 06:06:27	96
2	Group I comprehensive Coal mining team	Mr. Chen	2017-12-20 06:07:04	2017-12-20 06:07:31	3103

the type of work, the name of the miner, the initial reporting time, and the last reporting time. The latitude and longitude of the miner can be determined according to the location number.

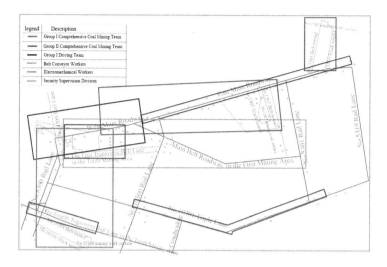

Fig. 2. Semantic schematic of some coal miners

As shown in Fig. 2, a coal mining area displays the semantic information of six kinds of coal miners. The main function of the security supervision division is to monitor the production safety in the mine, so the miners' trajectories are all over the coal mine. The main function of the belt conveyor workers is to control the belt conveyor, which determines that their working area is fixed, and their moving path is limited. The driving team is the roadway excavation team, so there are more stopping points in their trajectories than in other trajectories. The function of the comprehensive coal mining team is to mine coal in a specific working surface, so their trajectories will stay in the working surface area for a long time. Considering the semantic information of underground workers, it is necessary to classify the trajectories of the same type of work according to the miners' job information before trajectory processing so that the trajectory information mining is the most accurate and most suitable for the actual situation.

2.2   Detection of Inflection Point

The inflection point is the turning angle of the adjacent trajectory segments, which can reflect a trend in the moving object trajectory [1]. The working area in the mine is relatively fixed, and most of the roads are linearly distributed. When the moving object in the mine has turning or wandering behaviour in the working area, an inflection point of the corresponding angle is generated, so the inflection point in the trajectory can be regarded as an important position point. In Fig. 3, α is angle at the inflection point from trajectory P_3P_4 to trajectory P_5P_6, that is, the inflection point is the point P_4. It can be seen in the figure that the formula to calculate the angle is as shown in Formula 1.

$$\angle\alpha = \arccos\left(\frac{P_3P_4^2 + P_4P_5^2 - P_3P_5^2}{2 \times P_3P_4 \times P_4P_5}\right) \tag{1}$$

where

$$P_3P_4 = \sqrt{(x_3 - x_4)^2 + (y_3 - y_4)^2}$$

$$P_4P_5 = \sqrt{(x_4 - x_5)^2 + (y_4 - y_5)^2}$$

$$P_3P_5 = \sqrt{(x_3 - x_5)^2 + (y_3 - y_5)^2}$$

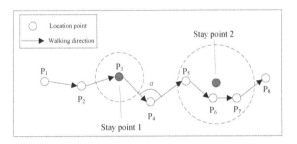

Fig. 3. Stop point diagram

Algorithm 1. Inflection Point Detection Algorithm

Input: Trajectory datasets.
Output: Inflection point set and non-inflection point set.
1: **for** every trajectory **do**
2:    **for** every point **do**
3:       calculate the turning angle α;
4:       **if** $\alpha >$ threshold **then**
5:          the point is inflection point;
6:       **else**
7:          the point is noise point or stay point;
8:       **end if**
9:    **end for**
10: **end for**

The screening formula for the inflection point is shown in Formula 2, where β is the angle threshold. When the inflection point of point P_i is less than or equal to β, it indicates that the moving object has changed its original trajectory at this point, so the point is marked as an important position point. When the angle is greater than β, it means that the moving object has a tendency to maintain its original motion trajectory, so it is necessary to further study the point to determine whether the point is a noise point or a stay point. Through many experiments, a better effect can be obtained by setting β to $110°$.

The time complexity for calculating the inflection point is $O(n)$, where n represents the number of location points. The Algorithm 1 is a pseudo code for inflection point detection.

$$P_i = \begin{cases} \text{an important position point,} & \alpha \leq \beta \\ \text{a noise point or a stay point,} & \alpha > \beta \end{cases} \tag{2}$$

2.3   Detection of a Stay Point

A stay point is a geographical point where the moving object stays for a long time. It may be a work area or a rest area in the mine. Stay points can be divided into two cases. The first is that the moving object stays in the same position within a certain time range. As shown in Fig. 3, the time that the moving object stays at point P_3 exceeds the time threshold, so P_3 is a stay point. The second case is that in a certain time range, the moving object wanders in a specific space [4]. As illustrated in Fig. 3, the moving object moves at P_5, P_6 and P_7 for greater than the time threshold, but the moving distance is less than the distance threshold. Therefore, the stay point 2 calculated by the average value of points P_5, P_6 and P_7 indicates the staying behaviour.

Algorithm 2. Stay Point Detection Algorithm

Input: Non-inflection point datasets.
Output: Candidate key sequence set.
 1: **for** every point **do**
 2:     calculate the time difference between adjacent points;
 3:     **if** time difference > time threshold **then**
 4:         calculate the distance difference between adjacent points;
 5:         **if** distance difference < distance threshold **then**
 6:             calculate the average value to add to the key sequence;
 7:         **end if**
 8:         add to the key sequence;
 9:     **end if**
10: **end for**

The stay point detection method in Algorithm 2 is based on the heuristic judgement of the distance threshold and the time threshold. First, it is determined whether the time difference between point P_1 and point P_2 is greater than the time threshold. If the condition is satisfied, it is determined whether the distance between the two points is less than the distance threshold. If the distance is less than the distance threshold, the requested stay point is the first case, and point P_2 is added to the key region sequence. Otherwise, it indicates that the requested stay point is the second case, and the two points P_1 and P_2 are added to the residence point sequence. The above process is repeated to determine whether the location point should be added to the residence point sequence until the distance threshold or time threshold is not satisfied. The average of the latitude and longitude of the location points in the residence point sequence is used to indicate the position of the stay point. The selected stay points are added to the key region sequence, and the other points are marked as noise points. Through several experiments, this paper set the distance threshold 50 m and the time threshold to 20 min.

3   Results and Analysis

The experimental data includes three months of underground personnel data, including the group I comprehensive mining team, the group II comprehensive mining team, the group I mechanized excavation team, and the group I tunneling team, with a total of 126,687 track points.

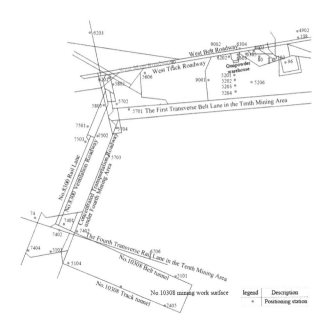

Fig. 4. Activity map of the group I comprehensive coal mining area

Figure 4 shows the specific activity map of the group I comprehensive coal mining team surrounded by the red frame in Fig. 2. The orange dots in the figure indicate the positioning stations. Each positioning station has its own number to indicate the position information of the moving object.

Using the above positioning data to conduct experiments, because the TS_HS algorithm does not process the data set, the effect of using the KLD method and not using the KLD method on the size of the data set is compared. Comparing the size of the original dataset with the size of the dataset after processing by the KLD method is shown in Fig. 5.

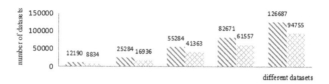

Fig. 5. Comparison of the quantity of raw data with the quantity of data processed by KLD

It can be clearly seen from Fig. 5 that experiments are performed on different data sets, and it is verified that the data sets processed using the KLD algorithm is reduced by approximately 1/4 of the track points compared to the original data sets. The reason is that we filter out the trajectory points that satisfy the inflection points and the stop points, and delete meaningless and non-semantic points, thereby reconstructing the original trajectory, and reducing data redundancy and computational complexity. In addition, in the subsequent cluster analysis work, DBSCAN is a data-sensitive analysis method, so the reduction of data quantity can greatly improve the analysis speed of DBSCAN algorithm.

4   Conclusions

For massive positioning information, it is particularly important to be able to quickly and effectively mine effective information, especially algorithms that are sensitive to data volume. Therefore, this paper proposes to perform key position screening on the edge side, namely KLD algorithm. The algorithm uses inflection points and the stay points for trajectory reconstruction and simplifies approximately 1/4 of the anchor points. It can reduce calculating time and convert the positioning data into a sequence of key positions with specific semantics.

References

1. Chang, C., Zhou, B.: Multi-granularity visualization of trajectory clusters using sub-trajectory clustering. In: 2009 IEEE International Conference on Data Mining Workshops, pp. 577–582. IEEE (2009)
2. Sun, X., Zhang, P., Chen, Y., Shi, L.: Interval multi-objective evolutionary algorithm with hybrid rankings and application in RFID location of underground mine. Control Decis. **32**, 31–38 (2017)
3. Wen, R., Tong, M., Tang, S.: Application of bluetooth communication in mine environment detection vehicle. In: 2017 7th IEEE International Conference on Electronics Information and Emergency Communication (ICEIEC), pp. 236–239. IEEE (2017)
4. Xie, R., Ji, Y., Yue, Y., Zuo, X.: Mining individual mobility patterns from mobile phone data. In: Proceedings of the 2011 International Workshop on Trajectory Data Mining and Analysis, pp. 37–44 (2011)

5. Zhang, A.L.: Research on the architecture of internet of things applied in coal mine. In: 2016 International Conference on Information System and Artificial Intelligence (ISAI), pp. 21–23. IEEE (2016)
6. Zhang, B., Tang, S., Jin, M., Xu, C., Tong, M.: Research on mine robot positioning based on weighted centroid method. In: 2018 International Conference on Robots & Intelligent System (ICRIS), pp. 17–20. IEEE (2018)

An Improved Spectral Clustering Algorithm Using Fast Dynamic Time Warping for Power Load Curve Analysis

Zhongqin Bi[1], Yabin Leng[1], Zhe Liu[2], Yongbin Li[3(✉)], and Stelios Fuentes[4]

[1] College of Computer Science and Technology,
Shanghai University of Electric Power, Shanghai, China
zqbi@shiep.edu.cn, lyb_0730@126.com
[2] State Grid Shanghai Electric Power Research Institute, Shanghai, China
liuzheacyy@163.com
[3] Office of Academic Affairs, Shanghai University of Electric Power, Shanghai, China
lybin40000@163.com
[4] Leicester University, Leicester, UK
stelios.fuentes@gmx.co.uk

Abstract. Cluster analysis of power loads can not only accurately extract the commonalities and characteristics of the loads, but also help to understand the users' habits and patterns of electricity consumption, so as to optimize the power dispatching and regulate the operation of the entire power grid. Based on the traditional clustering methods, this paper proposes a clustering algorithm that can automatically determine the optimal cluster number. Firstly, Fast-DTW algorithm is used as the similarity measuring function to calculate the similar matrix between two time series, and then Spectral Clustering and Affinity Propagation (AP) algorithm are used for clustering. It is combined with Euclidean distance, DTW and Fast-DTW algorithms to evaluate the algorithm effect. By analyzing the actual power data, our results show that the improved external performance evaluation index ARI, AMI and internal performance evaluation index SSE are significantly improved and have better time series similarity and accuracy. Applying the algorithm to more than six thousands of users, twelve kinds of typical power load patterns can be obtained. For any other load curve, it can be mapped to a standard load by feature extraction. The corresponding prediction model is adopted, which is of great significance to reduce the peak power consumption, adjust the electricity price appropriately and solve the problem of system balance.

Keywords: Cluster analysis · Time series · Fast-DTW · Spectral clustering

1 Introduction

With the continuous and steady development of the social economy, the power load has grown rapidly. In recent years, there has been a phenomenon of power

© ICST Institute for Computer Sciences, Social Informatics and Telecommunications Engineering 2020
Published by Springer Nature Switzerland AG 2020. All Rights Reserved
J. Liu et al. (Eds.): MobiCASE 2020, LNICST 341, pp. 143–159, 2020.
https://doi.org/10.1007/978-3-030-64214-3_10

shortage appears in many areas of China during the peak period of power consumption. At the same time, the traditional method of increasing investment in power generation is not economical enough. Thus, alleviating peak power shortages by exploiting demand-side resources has received more attention. At present, China's electricity market is not yet complete, and the demand-side management (DSM) mainly adopts extensive electricity consumption mode, lacking of serious consideration of load form and low satisfaction with electricity consumption. Therefore, the cluster analysis of power load test data is the cornerstone of DSM and even overall planning of the entire power system.

In order to further investigate the standardized model of power load curve, improve the accuracy of clustering analysis, and provide an effective scheme for the supply-demand side reform of power resource consumption, scholars have applied data mining algorithm to the analysis of power load curve. Familiar algorithms include K-means Algorithm, Fuzzy C-means Algorithm, Hierarchical Clustering Algorithm, and Self-Organizing Feature Map Network Algorithm. Ioannis et al. used two different methods to improve the K-means Algorithm and applied it to the time series analysis of power load curve. The results showed that the clustering accuracy was significantly improved [1]. Gao and Zhao et al. combined Fuzzy C-means Algorithm, Conjugate Gradient Algorithm and Deep Belief Network, and proposed a new combination model for short-term photovoltaic power load forecasting, which achieved ideal results as well [2]. The analysis of the electricity consumption patterns of residential users is helpful to improve the accuracy of load forecasting model and to provide reliable and high-quality electricity supply for electric power enterprises, which is of great practical significance to the reform of electricity price. Load forecasting plays an important role in the planning, dispatching, operation, maintenance and control of modern power system [3]. Moreover, the development of energy industry, the change of load demand and the popularization of smart electricity meters all need a new load model to support the research of power system, so a new method of random load modeling of smart electricity meters is proposed [4]. In addition, two data sets were compared by using clustering algorithm, the commonly used data set reduction techniques and feature extraction methods of load patterns were analyzed, and the existing research on power customer clustering was summarized, with emphasis on the main research results [5]. A new load consumption pattern clustering model is proposed to identify periods with similar load levels, typical load patterns of each customer, and periods at different load pattern levels, so as to provide guidance and suggestions for DSM strategies [6]. In addition, different types of load time series are transformed into mapping models to reduce interference and improve differentiated cluster efficiency of power customers [7].

Cluster analysis is an unsupervised learning method in data mining technology. This paper not only proposes a spectral clustering algorithm based on particle swarm, which improves text clustering, but also provides an effective solution for information retrieval, information extraction and document organization [8]. Furthermore, a method of spectral clustering based on iterative optimization is studied, which solves the problem of spectral decomposition of large-scale high-dimensional data sets and provides an effective solution for spectral clustering [9]. Fully preserving the information integrity of time series is a key link of power load curve clustering analysis. On this basis, an adaptive dynamic time structuring algorithm (ACDTW) is proposed to reasonably arrange the mapping points between two time series. On the one hand, it avoids excessive stretching and compression of time series; on the other hand, it solves the problem that the loss of key feature information will affect the classification accuracy [10]. Time series clustering is a key link in the process of power load curve analysis. It is difficult to fit the similarity of shape and contour of time series adequately. Although many literatures have provided an effective method to extract the standardized model of power load curve and obtained satisfactory results, there is still much room for improvement in the clustering analysis of time series.

Spectral clustering is an algorithm developed from graph theory. Compared with traditional clustering algorithm, it has the advantages of simple implementation and perfect clustering effect. However, the disadvantages are also obvious. Firstly, the clustering center cannot be automatically determined; secondly, the algorithm is prone to local optimization; thirdly, the similarity of time series shape and contour cannot be guaranteed. On this basis, a spectral clustering method without considering the internal characteristics of time series is proposed. Firstly, by comparing Euclidean distance, DTW and Fast-DTW, the influence of similarity measure on time series clustering is studied [11]. Secondly, the clustering effect of K-means Algorithm and AP Neighbor Propagation Algorithm on eigenvectors is compared. Finally, we propose a new Fast-DTW-AP Spectral Clustering Algorithm, which can not only automatically select the optimal number of clusters in arbitrary sample space, but also effectively avoid the algorithm falling into the phenomenon of local optimization. In addition, it has a better affinity for high-dimensional time series data or sparse data.

After the performance test of the algorithm on the standard data set and the real data set, we applied the proposed method to thousands of home power users, and cluster them into 12 clusters. According to the clustering results obtained, the characteristics of each type of load are analyzed, and the standardized model of load is established, which can be applied to different types of load. This article addresses three issues in the definition of a standardized model. The first is to propose a time series load clustering algorithm which is helpful to improve the prediction accuracy in high dimensional space. The second is to use the improved algorithm to cluster the load and then design a load prediction model suitable for this feature. The third is to adjust the electricity price according to the load prediction model, which is of great significance to the safety, economy and stable operation of the power system.

In this paper, a clustering algorithm combining the selection of internal similarity matrix in spectral clustering with AP neighbor propagation is proposed, which can automatically determine the clustering center and effectively avoid falling into local optimization. The Fast-DTW-AP improved spectral clustering algorithm is applied to thousands of households, and 12 standard power models are obtained. Through the extraction of users' consumption habits and patterns, accurately grasp the law of electricity consumption. On this basis, a more effective electricity price is designed to adjust the residents' demand.

The paper is organized as follows: The second sector is the model building. Sector 2.1 introduces three similarity measurement methods: Euclidean distance, DTW and fast-DTW. Sector 2.2 improves the K-means algorithm of spectral clustering, making the clustering effect of the algorithm more excellent. In the Sect. 3, internal and external indexes are used to evaluate the experiment, and the experimental results of the Fast-DTW-AP improved spectral clustering and other clustering algorithms in time series data sets and standard data sets are analyzed and compared.

2   Model Building

Time series of power demand is the key information source of consumer behavior. Although some scholars have studied the load pattern of extracting a large number of power users, there are few researches about calculation of time series. Therefore, improving traditional clustering techniques, optimizing the number of clustering, and improving the quality of clustering and the similarity of time series have become an important topic.

2.1   Similarity Measure

The core of improved power time series clustering analysis is the similarity measure that constructs the similarity matrix between two power time series curves. In order to study the role of similarity measure in power time series clustering analysis, Euclidean Distance, DTW and Fast DTW are used as similarity measures in the application of spectral clustering, and the final clustering results are analyzed and compared reasonably.

Euclidean Distance. For two power time series curves U and V with length $|U|$ and $|V|$ respectively.

$$U = \{U_1, U_2, \ldots, U_{|U|}\} \tag{1}$$

$$V = \{V_1, V_2, \ldots, V_{|V|}\} \tag{2}$$

The Euclidean distance requires that the sample power time series curve must be equal in length, that is, $|U| = |V|$. The formula for defining the distance between U and V in n-dimensional space is:

$$ED(U, V) = \sqrt{\sum_{i=1}^{n}(u_i - v_i)^2} \tag{3}$$

Euclidean distance is the most commonly used distance measurement method. It measures the absolute distance between two power series curves,

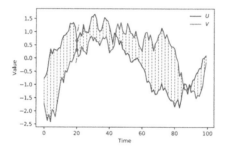

Fig. 1. Euclidean distance between two power time series curves

but it can only measure the time series of the same length. However, the power time series curve generated by power load in the actual power generation process is disordered, so Euclidean distance is difficult to predict whether there is a similar trend between the two power time series. As shown in Fig. 1, the local peaks of the curves and of the power time series do not match, which is caused by the fact that the Euclidean distance can only match the two series point-to-point.

Dynamic Time Warping Algorithm (DTW). Dynamic time warping algorithm is a non-linear measure of the minimum distance between two power time series curves [12]. Its purpose is to find the sum of the minimum cumulative distance of all corresponding points of two power time series curves, namely, to find the shortest integration path. It represents the optimal matching between two power time series curves, fully guarantees the shape and contour similarity of the two power time series curves and breaks through the limitation of Euclidean distance for the calculation of unequal length power time series curves.

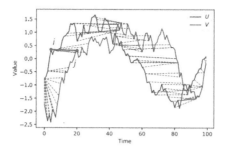

Fig. 2. DTW nonlinear measurement between two power time series curves

As can be seen from Fig. 2, point i in the power time series curve U can be well aligned with point $j(i = j)$ in the series V. The curve integration path start from the start point $(1, 1)$ corresponding to the two sequences and ends at the end point $(|U|, |V|)$ corresponding to the two sequences. For the cumulative distance $Dist(i, j)$ is defined as:

$$Dist(i, j) = min\{Dist(i - 1, j - 1), Dist(i - 1, j), Dist(i, j - 1)\} + d(u_i, v_j) \quad (4)$$

$Dist(1, 1) = d(1, 1)$, and $d(u_i, v_j)$ are usually calculated using Euclidean distance. The optimal regular path is determined by $Dist$ $(|U|, |V|)$. The DTW algorithm can automatically match the peaks and is not limited by the length of the template. It is suitable for clustering analysis of power curve.

However, the DTW algorithm also has obvious defects in the actual power curve analysis, that is, the algorithm complexity is too high. When two power time series curves are relatively long, the efficiency will be slow and the regularity will be too large, which may easily lead to the wrong matching of the power curve.

Fast Dynamic Time Warping Algorithm (Fast-DTW). The Fast Dynamic Time Warping Algorithm is an acceleration algorithm of the classical dynamic time structuring algorithm. The algorithm combines limited search space and data abstraction. On the basis of fully ensuring the accuracy of the algorithm, a reasonable solution is provided for the clustering analysis of power curves with large amount of data in practical application.

The three steps of the Fast-DTW algorithm are as follows: First, the original power time series curve is extracted with coarse-grained data, and repeated iterative optimization is performed. Where, the coarse-grained data points are the average values of the corresponding fine-grained data points. Second, the DTW algorithm is run granularity on coarse-grained power time series curves. Finally, the regular path obtained on the coarser granularity is further fine-grained into a finer-grained power time series curve through a grid [13].

Table 1. The comparison of similarity measures

Algorithm	Complexity	Alignment
ED	O(N)	One-to-one
DTW	O(N^2)	One-to-many
Fast-DTW	O(N)	One-to-many

The algorithm complexity of the three similarity measures is shown in Table 1. The algorithm complexity of the Euclidean distance is N, but it can only meet the requirements of point-to-point, that is, it can only measure power time series curve of equal length. DTW can match different power time series curve, but the algorithm complexity is high. The Fast-DTW algorithm not only meets the needs of unequal length power time series curve, but also reduces the complexity of the algorithm.

2.2   Clustering Algorithms

The key of power load curve clustering analysis is the choice of clustering algorithm. Because of its excellent clustering effect, spectral clustering algorithm is more and more widely used. The Affinity Propagation Algorithm can automatically determine the clustering center, which is not sensitive to the initial power data, and can complete the clustering of large-scale and multi category data sets in a short time, which is suitable for the clustering analysis of power time series curves with large amount of data. In order to obtain more ideal clustering results, this paper makes appropriate improvements based on the two clustering algorithms.

Spectral Clustering. The main idea of spectral clustering is to transform all the power data into points in space. These points can be connected by edges. The edge weight value between the two points with a long distance is lower, while the edge weight value between the two points with a short distance is higher. By cutting the graph composed of all data points, the edge weight between different subgraphs after cutting is as low as possible, and the edge weight sum within the subgraph is as high as possible, so as to achieve the purpose of clustering [14].

There are obvious disadvantages when using original spectral clustering for cluster analysis: (1) The cluster center cannot be determined automatically; (2) The final clustering effect is largely affected by the similarity matrix and feature vector clustering algorithm; (3) It is very sensitive to the choice of clustering parameters.

Affinity Propagation (AP) Algorithm. In spectrum clustering, K-means algorithm is used to cluster the eigenvector space. However, K-means algorithm is very sensitive to the selection of the initial clustering center, and its hill-climbing optimization algorithm often fails to obtain the global optimal solution, so the Affinity Propagation (AP) algorithm is introduced. The AP algorithm is a clustering algorithm based on "information transfer" between data points [15]. The algorithm does not need to determine the number of clusters before running it. In addition, because the actual points in the data set are selected, the clustering effect is better with the cluster center as the representative of each class. The basic steps are as follows:

(1) Euclidean distance is used to calculate the similarity between two data points, and a similarity matrix S is constructed. S is a $n \times n$ matrix.
(2) Calculate the attraction matrix:

$$R_{t+1}(i, k) = (1 - \lambda) \bullet R_{t+1}(i, k) + \lambda \bullet R_t(i, k) \tag{5}$$

$$R_{t+1}(i, k) = S(i, j) - max\{A_t(i, j) + R_t(i, j)\} \tag{6}$$

Among them, $S(i, j)$ is an element in the similarity matrix S, which indicates the ability of point j to be the clustering center of point i. Generally, a negative

Euclidean distance is used. The larger $S(i, j)$, The closer the distance between the points, the higher the similarity. The degree of attraction $R(i, k)$ indicates the degree that point K is suitable to be the clustering center of data point I, which is a process of selecting point K from point I. The damping coefficient λ is used for the convergence of the algorithm, and the value range is $[0.5, 1]$.

(3) Calculate the membership matrix:

$$A_{t+1}(i, k) = (1 - \lambda) \bullet A_{t+1}(i, k) + \lambda \bullet A_t(i, k) \tag{7}$$

$$A_{t+1}(i, k) = min\{0, R_{t+1}(k, k) + \sum_{j \in i, k} max\{0, R_{t+1}(j, k)\}\} \tag{8}$$

The degree of belonging $A(i, k)$ indicates the suitability of point i to select point k as its clustering center. This is a process in which point k selects point i. $A_{t+1}(i, k)$ represents the new $A(i, k)$, and $A_t(i, k)$ represents the old $A(i, k)$.

(4) Iteratively update R and A values.

The selection of the appropriate clustering center is crucial to the quality of the final clustering effect. When the value of $A(i, k) + R(i, k)$ is larger, it means that the probability of K points as the cluster center is greater. In order to find the maximum value, it is necessary to iteratively update the R and A values to obtain the most suitable cluster center. The termination condition of iteration is that the cluster center is not updated to a certain extent or reaches the maximum number of iterations (generally 15 times). After getting the most suitable cluster center, the data set can be classified directly.

Improved Spectral Clustering Algorithm. In order to extract the standardized model of power time series accurately and effectively, more reliable clustering analysis results are obtained. In this paper, a Fast-DTW-AP spectral clustering algorithm (F-A-S) is proposed based on the original spectral clustering. As long as the data points are entered, the cluster center and the number of clusters can be automatically determined. By analyzing the number of clusters, the inherent shortcomings of spectral clustering algorithm are solved, which makes the algorithm not only suitable for arbitrary shape sample space clustering, but also can effectively prevent the algorithm from falling into the local optimal phenomenon. In addition, the improved algorithm has better performance in processing unequal length power time series curves, which fully guarantees the similarity of shape and contour of power time series curves, and its performance is far better than that in processing high-dimensional data and sparse data clustering.

Through comparative experiments, the traditional clustering algorithm finally found that the improved algorithm significantly improved the clustering effect of time series, and the external evaluation indexes ARI, AMI and internal evaluation indexes SSE of the clustering were significantly improved. The specific algorithm steps are as follows:

1. Enter a power time series data set and use the Fast-DTW algorithm to calculate the similarity between each data point.

$$S(i, j) = FastDTW(i, j) \tag{9}$$

The similarity matrix S is generated after calculating the similarity between the data points by using the Fast-DTW distance. S is a $n \times n$ matrix composed of $S(i, j)$. The value of element p in the matrix is 1. The purpose is to preserve the integrity of its own information to the greatest extent.

$$S = \begin{bmatrix} p & S(1,2) & \cdots & S(1,n) \\ S(2,1) & p & \cdots & S(2,n) \\ \vdots & \vdots & \ddots & \vdots \\ eS(n,1) & S(n,2) & \cdots & p \end{bmatrix} \tag{10}$$

2. Apply the following formula to calculate the degree matrix.

$$d_i = \sum_{j=1}^{n} S_{ij} \tag{11}$$

A degree matrix D can be constructed from the similarity matrix. The degree matrix D is a $n \times n$ diagonal matrix composed of d_i. Each element on the diagonal is the sum of the elements of each row of the corresponding similarity matrix, and all other elements are 0.

3. Laplace matrix L is calculated according to similarity matrix S and degree matrix D.

$$L = D - S \tag{12}$$

The degree matrix D and the similarity matrix S are different to generate a Laplacian matrix L. In order to obtain better clustering results, after obtaining the Laplacian matrix, this paper uses a symmetric normalization method to normalize the Laplacian matrix.

$$L = D^{-1/2}(D - S)D^{-1/2} \tag{13}$$

4. The eigenvalues of Laplace matrix L are calculated, the eigenvalues are sorted from small to large, and the eigenvector $u_1, u_2, \ldots, u_n$ corresponding to each eigenvalue is calculated. The matrix $U = \{u_1, u_2, \ldots, u_n\}$ (n rows and n columns) is composed of n column vectors.

5. Let Y_i be the i-th row vector of $U(i = 1, 2, \ldots, n)$, and then form a new matrix $Y = \{y_1, y_2, \ldots, y_n\}$.

6. Use the Affinity Propagation (AP) algorithm to cluster the new sample point $Y = \{y_1, y_2, \ldots, y_n\}$ and divide it into k clusters.

Algorithm 1. The Fast-DTW-AP Spectral Clustering Algorithm

Input:
    Sample dataset $X = \{x_1, x_2, \ldots, x_n\}$;
Output:
    Cluster set $C = \{C_1, C_2, \ldots, C_k\}$;
1: Calculate the Fast-DTW distance of X, and obtain the similarity matrix S;
2: Compute the standardized laplacian using Eq. (13);
3: Compute the eigen vectors $U_1, U_2, \ldots, U_n$ of Laplacian;
4: Let $X \in R^{n \times n}$ which contains the vectors $U_1, U_2, \ldots, U_n$ as a column;
5: Let $Y \in R^{n \times n}$ be the vector corresponding to the ith row of U;
6: Group the points $Y_i \in R^{n \times n}$ with the AP clustering algorithm into $\{C_1, C_2, \ldots, C_k\}$

3   Experiment and Analysis

3.1   Basic Description of the Experimental Environment

All the experiments in this paper are carried out on a computer equipped with windows 10 operating system, Intel (R) core (TM) i5-3230m 2.60 GHz CPU and 8 GB RAM, and the algorithm is implemented with Python 3.7 software.

The Comprehensive Control Chart Time Series (SCCTS) data set used in this paper is a standard time series test data set in the UCI database. As shown in Table 2, the data set contains 6 different classes with a total of 600 rows and 60 columns [16].

Table 2. Classification and abbreviations of SCCTS

Number	Abbreviation	Class
0–99	N	Normal
100–199	C	Cyclic
200–299	I	Increasing trend
300–399	D	Decreasing trend
400–499	US	Upward shift
500–599	DS	Downward shift

The real power consumption data comes from the smart electricity customer behavior test conducted by Energy Regulatory Commission (CER) during 2009–2010. Smart meters measure consumption in KW per half hour. The data set contains more than 6000 customer records, of which 66% are residents, 7% are small and medium-sized enterprises, and 27% are other customers [17].

3.2   Clustering Quality Evaluation Index

The clustering quality evaluation index consists external evaluation index and internal evaluation index. Among them, the external evaluation index is to apply the clustering algorithm to the standard test data set with clear classification, and then calculate the clustering accuracy of the algorithm to the data set with relevant indexes. While the internal indicators refer to pre-defined evaluation criteria, which are usually used to describe some intrinsic characteristics and quantitative values of clustering after clustering in order to evaluate the quality of clustering results.

External Evaluation Indicators. The adjusted clustering method was evaluated by using the Adjusted Mutual Information (AMI) [18] and Adjusted Rand index (ARI) [19] as external standards for evaluating the quality of clustering. Given two sets $U = \{U_1, U_2, \ldots, U_i\}$ and $V = \{V_1, V_2, \ldots, V_j\}$ i is the number of clusters in U, and j is the number of clusters in V. ARI represents the number of paired samples belonging to the same classification or different classifications in two sets, and the expected value between U and V is defined as:

$$E(U,V) = \left| \sum_{u_i} \binom{n_i}{2} \sum_{v_j} \binom{n_j}{2} \right| / \binom{n}{2} \tag{14}$$

ARI is defined as:

$$ARI(U,V) = \frac{\sum_{u_i} \sum_{v_j} \binom{n_{ij}}{2} - E(U,V)}{\frac{1}{2}[\sum_{u_i} \binom{n_i}{2} + \sum_{v_j} \binom{n_j}{2}] - E(U,V)} \tag{15}$$

AMI is a clustering evaluation index based on the degree of correlation between two random variables, called mutual information MI, that is, the amount of information about one random variable contained in one random variable. AMI is defined as:

$$AMI(U,V) = \frac{MI(U,V) - E\{MI(U,V)\}}{max\{H(U), H(V)\} - E\{MI(U,V)\}} \tag{16}$$

Both the AMI and ARI indexes have a value range of $[0, 1]$, and the larger the value, the more consistent the clusters divided by the standard clusters.

Internal Evaluation Indicators. In order to better evaluate the clustering effect of the algorithm, The sum of squares due to error (SSE) is used as the internal standard to evaluate the clustering quality.

$$I_{SSE} = \sum_{i=1}^{k} \sum_{x \in G_i} \|x - o_i\|^2 \tag{17}$$

SSE is the sum of the squares of the distances from the data points in all subclasses to the corresponding cluster centers after clustering.

3.3 Experiment Analysis

By comparing the performance of standard test data and actual power consumption, the validity and effectiveness of the proposed algorithm are proved. First, in order to evaluate the effectiveness of the proposed algorithm, the Fast-DTW-AP spectral clustering algorithm (F-A-S) is used for clustering on the standard SCCTS data set, and the external evaluation standards ARI, AMI and internal evaluation standard SSE are used to measure the final result. And the clustering effect is compared with other improved spectral clustering algorithms based on DTW and AP, such as the original spectral clustering algorithm (S), Fast-DTW spectral clustering algorithm (F-K-S), DTW-AP spectral clustering algorithm (D-A-S), etc.

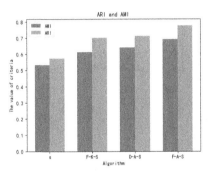

Fig. 3. External evaluation indicators of the Fast-DTW-AP spectral clustering algorithm and other improved spectral clustering algorithms

As can be seen from the comparison results in Fig. 3, the improved Fast-DTW-AP spectral clustering algorithm (F-A-S) has significantly improved the performance of AMI and ARI indexes, indicating that the algorithm proposed in this paper has higher fitting accuracy for time series and can better realize the classification of standard data. As Euclidean distance was used as the similarity measure in the original spectral clustering, it could not match the unequal time series. Therefore, compared with the traditional spectral clustering algorithm, the AMI and ARI evaluation indexes of the Fast DTW-AP spectral clustering algorithm were improved by 16.2% and 18.4%, respectively. Since AP algorithm has better processing effect than K-means algorithm in dealing with complex time series, compared with Fast DTW and K-means (F-K-S) combined algorithm, AMI and ARI of Fast-DTW-AP spectral clustering algorithm are improved by 7.1% and 8.6% respectively. Compared with combination algorithm of DTW algorithm and AP algorithm (D-A-S), AMI and ARI of Fast-DTW-AP spectral clustering algorithm increased respectively 6.4% and 7.3%, which is because fast DTW solves the problem of excessive regularity of DTW. The above three comparative experiments fully show that our proposed algorithm has better clustering effect.

Table 3. Internal evaluation index of Fast-DTW-AP

Algorithm	SSE
S	1.0833×10^5
F-K-S	1.0422×10^5
D-A-S	1.0359×10^5
F-A-S	1.0297×10^5
Standard time series set	1.0195×10^5

As can be seen from Table 3, the SSE index value of the improved Fast-DTW-AP spectral clustering algorithm (F-A-S) is lower than the other three comparison algorithms, and the intra cluster variance is closer to 1, which indicates that the improved Fast-DTW-AP spectral clustering algorithm has achieved excellent clustering results. At the same time, it is noted that the intra class variance of this algorithm is closer to the result of standard data. Experimental results show that the proposed algorithm has better performance in processing time series.

In order to further verify the feasibility of the algorithm, K-Medoids algorithm and AP algorithm were used to cluster the same data set, and AMI and ARI evaluation indexes were used to evaluate the final clustering effect.

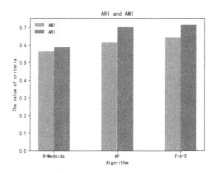

Fig. 4. Evaluation indicators of K-Medoids, AP algorithm, and improved spectral clustering algorithm

The experimental results are shown in Fig. 4. Compared with the improved k-Medoids algorithm, AMI and ARI are improved by 8.2% and 13.8% respectively. Compared with the AP algorithm, the improved spectral clustering The algorithm increased by 6.8% and 3.3% respectively, which shows that the proposed algorithm also performs well when compared with some other data mining algorithms.

In order to make the effectiveness of the algorithm more rigorous and ensure that it not only has a better clustering effect on this data set, the staff absence standard time series data set (AAW) was selected from the UCI database to conduct clustering effect test. The data set is derived from real data and contains a total of 740 instances and 21 attributes [20].

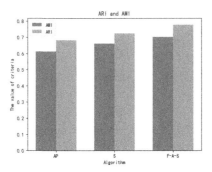

Fig. 5. Evaluation indicators of spectral clustering, AP, and improved spectral clustering algorithm in AAW dataset

As shown in Fig. 5, after replacing the AAW data set, the clustering effect obtained by using the improved spectral clustering algorithm also has obvious advantages. Compared to the AP algorithm, the improved spectral clustering algorithms AMI and ARI The index is increased by 8.8% and 9.7%, respectively. Compared with the traditional spectral clustering algorithm, the AMI and ARI indexes of the improved spectral clustering algorithm are improved by 3.9% and 5.4% respectively, indicating that the improved spectral clustering algorithm has better performance. The same performance is achieved when applied to other data sets.

Finally, the actual efficiency of the algorithm is verified by using actual energy consumption data from THE Energy Council (CER). In this paper, the power data set is preprocessed, including deleting users who lose data, deleting data that is not suitable for analysis near zero, and using minimum-maximum normalization to map the data uniformly to the interval [0, 1].

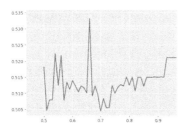 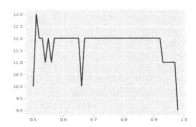

Fig. 6. Relationship between the number of clusters and the contour coefficient

On the basis of data processing, we apply the Fast-DTW-AP spectral clustering algorithm to cluster the power load curves of thousands of users' power consumption patterns. Figure 6 shows the process of automatically determining the optimal number of clusters in the analysis, and the most accurate cluster number is obtained by calculating the optimal damping coefficient. Therefore, the contour coefficient is used to evaluate the clustering effect, and the damping coefficient is taken as a parameter, whose variation range is between (0.5, 1). The relationship between the damping coefficient and the profile coefficient is obtained by taking the damping coefficient as the horizontal axis and the profile coefficient as the vertical axis. The higher the contour coefficient is, the better the clustering effect of the corresponding damping coefficient is. It is of great significance to find the optimal damping coefficient for the final clustering effect. As shown in Fig. 6, the highest contour coefficient $Y = 0.533382$ is obtained when $X = 0.658032$, indicating that the optimal damping coefficient of the data set is 0.658032, and the optimal number of clusters is 12. In order to achieve the best clustering effect, the optimal number of clusters is obtained by calculating the optimal damping coefficient.

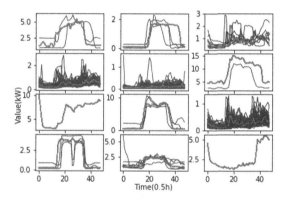

Fig. 7. Clustering effect of CER dataset (Color figure online)

Finally, 12 types of power load curve standardized models were obtained. For each clustering result, a power load curve is drawn as Fig. 7, and a typical load curve is extracted by taking time as the abscissa and electricity load power as the ordinate. In this way, the electricity consumption curve (blue line) and typical load curve (red line) of each household can be obtained, so as to accurately extract the commonness and difference of load.

4   Conclusion

In order to effectively extract valuable information from power data, optimize power dispatch and regulate the operation of the entire power grid, this paper proposes a Fast-DTW-AP improved spectral clustering algorithm based on time series. The main contributions are summarized as follows: First, the external index of AMI, ARI and internal index of SSE were used to evaluate the clustering results of the Fast-DTW-AP improved spectral clustering algorithm and other three time series clustering methods. The spectral clustering algorithm can effectively retain the morphological and contour similarities between time series.

Second, comparing Fast-DTW-AP improved spectral clustering algorithm with other two commonly used data mining algorithms, we found that the external evaluation indexes AMI and ARI were significantly improved, which further proved the robustness and practical feasibility of the algorithm.

Third, the Fast-DTW-AP spectral clustering algorithm was tested on SCCTS, AAW and CER of Irish smart meter. Multiple experimental results show that the Fast-DTW-AP improved spectral clustering algorithm has achieved the best performance.

Fast-DTW-AP spectral clustering algorithm has certain advantages compared to other clustering algorithms when processing time series. In general, a reasonable power model is designed to help adjust the appropriate electricity price, minimize peak power consumption, and solve the problem of system balance.

Acknowledgment. This work is supported by the National Nature Science Foundation of China (No. 61972357, No. 61672337).

References

1. Panapakidis, I.P., Christoforidis, G.C.: Implementation of modified versions of the K-means algorithm in power load curves profiling. Sustain. Cities Soc. **35**, 83–93 (2017)
2. Gao, Z., Li, Z., Bao, S.: Short term prediction of photovoltaic power based on FCM and CG DBN combination. J. Electr. Eng. Technol. **15**, 333–341 (2020)
3. Fu, X., Zeng, X.J., Feng, P., Cai, X.: Clustering-based short-term load forecasting for residential electricity under the increasing-block pricing tariffs in China. Energy **165**, 76–89 (2018)
4. Khan, Z.A., Jayaweera, D., Alvarez-Alvarado, M.S.: A novel approach for load profiling in smart power grids using smart meter data. Electr. Power Syst. Res. **165**, 191–198 (2018)
5. Rajabi, A., Eskandari, M., Ghadi, M.J., Li, L., Zhang, J., Siano, P.: A comparative study of clustering techniques for electrical load pattern segmentation. Renew. Sustain. Energy Rev. **120**, 109628 (2019)
6. Charwand, M., Gitizadeh, M., Siano, P., Chicco, G., Moshavash, Z.: Clustering of electrical load patterns and time periods using uncertainty-based multi-level amplitude thresholding. Int. J. Electr. Power Energy Syst. **117**, 105624 (2020)
7. Motlagh, O., Berry, A., O'Neil, L.: Clustering of residential electricity customers using load time series. Energy **237**, 11–24 (2019)
8. Janani, R., Vijayarani, S.: Text document clustering using spectral clustering algorithm with particle swarm optimization. Expert Syst. Appl. **134**, 192–200 (2019)
9. Zhao, Y., Yuan, Y., Nie, F., Wang, Q.: Spectral clustering based on iterative optimization for large-scale and high-dimensional data. Neurocomputing **318**, 227–235 (2018)

10. Wan, Y., Chen, X.-L., Shi, Y.: Adaptive cost dynamic time warping distance in time series analysis for classification. J. Comput. Appl. Math. **319**, 514–520 (2017)
11. Han, T., Peng, Q., Zhu, Z., Shen, Y., Huang, H., Abid, N.N.: A pattern representation of stock time series based on DTW. Phys. A Stat. Mech. Appl. **550**, 124161 (2020)
12. Kang, Z., et al.: Multi-graph fusion for multi-view spectral clustering. Knowl.-Based Syst. **189**, 105102 (2019)
13. Salvadora, S., Chan, P.: Toward accurate dynamic time warping in linear time and space. Intell. Data Anal. **11**, 561–580 (2007)
14. Cao, Y., Rakhilin, N., Gordon, P.H., Shen, X., Kan, E.C.: A real-time spike classification method based on dynamic time warping for extracellular enteric neural recording with large waveform variability. J. Neurosci. Methods **261**, 97–109 (2016)
15. Han, Y., Wu, H., Jia, M., Geng, Z., Zhong, Y.: Production capacity analysis and energy optimization of complex petrochemical industries using novel extreme learning machine integrating affinity propagation. Energy Convers. Manag. **180**, 240–249 (2019)
16. Alcock, R.: Synthetic control chart time series data set (1999). http://archive.ics.uci.edu/ml/machine-learning-databases/synthetic_control-mld/. Accessed via UCI
17. CER smart metering project-electricity customer behaviour trial (2017). http://www.ucd.ie/issda/data/commissionforenergyregulationcer/. Accessed via the Irish Social Science Data Archive
18. Xie, J., Gao, H., Xie, W.: Robust clustering by detecting density peaks and assigning points based on fuzzy weighted K-nearest neighbors. Inf. Sci. **354**, 19–40 (2016)
19. Xie, J., Zhou, Y., Ding, L.: Local standard deviation spectral clustering. In: IEEE International Conference on Big Data and Smart Computing, vol. 143, pp. 242–250 (2018)
20. Martiniano, A., Ferreira, R.P., Sassi, R.J.: Absenteeism at work Data Set (2010). http://archive.ics.uci.edu/ml/datasets/Absenteeismatwork/. Accessed via UCI

AI Application

Research on Text Sentiment Analysis Based on Attention C_MGU

Diangang Wang[1], Lin Huang[1], Xiaopeng Lu[2(✉)], Yan Gong[1], and Linfeng Chen[3]

[1] State Grid Sichuan Information and Communication Company, Chengdu 610041, Sichuan, China

[2] School of Computer Science and Technology, Shanghai University of Electric Power, Shanghai 200090, China
2217297003@qq.com

[3] School of Information and Electronic Engineering, Zhejiang University of Science and Technology, Hangzhou 310023, Zhejiang, China

Abstract. Combining the advantages of the convolutional neural network CNN and the minimum gated unit MGU, the attention mechanism is merged to propose an attention C_MGU neural network model. The preliminary feature representation of the extracted text is captured by the CNN's convolution layer module. The Attention mechanism and the MGU module are used to enhance and optimize the key information of the preliminary feature representation of the text. The deep feature representation of the generated text is input to the Softmax layer for regression processing. The sentiment classification experiments on the public data sets IMBD and Sentiment140 show that the new model strengthens the understanding of the sentence meaning of the text, can further learn the sequence-related features, and effectively improve the accuracy of sentiment classification.

Keywords: Sentiment analysis · C_MGU · Attention mechanism

1 Introduction

With the widespread application of the Internet, the Internet (such as blogs, forums, and social service networks such as Douban, Public Comment) has generated a large number of user-engaged comment information with emotional factors such as people, events, and products. Recording various life states in the form of text to express emotions and attitudes has occupied an increasingly important position in people's daily communication. As a carrier of emotion, through the study of network text information, it is possible to analyze the emotional changes of network users, understand the selection preferences of network users from multiple aspects, and better understand the behaviors of network users. In addition, the government can supervise and manage the network,

Science and technology projects funded by State Grid Sichuan Electric Power Company (NO.: 521947140005).

J. Liu et al. (Eds.): MobiCASE 2020, LNICST 341, pp. 163–173, 2020.
https://doi.org/10.1007/978-3-030-64214-3_11

maintain Internet order, and use the power of public opinion to guide network users to form correct values and prevent network deviations from extending into real life.

Emotion is the advanced behavior of human intelligence, and people express emotions in a variety of ways. In order to understand the emotions in the text, the emotions need to be classified. The purpose is to classify the emotions into positive, negative emotions or more detailed emotion categories [1]. Text sentiment analysis is also known as opinion mining, which uses natural language processing, text analysis, and other methods to analyze, process, reason, and summarize texts with emotions [2, 3]. Existing sentiment analysis methods are mainly divided into two categories based on sentiment dictionary matching methods and machine learning-based methods.

As deep learning has gradually become a research hotspot in the field of natural language processing, the technology of using deep learning methods based on sentiment dictionary matching to solve sentiment analysis problems has also developed rapidly. Many scholars have optimized the sentiment characteristics of text sentiment analysis [4]. Zhang et al. [5] proposed a strategy based on sentiment dictionary, which successfully classified sentiment of online user-generated text. Zhang et al. [6] proposed a sentiment dictionary-based sentiment analysis method for Chinese Weibo texts. The sentiment values of Weibo texts were obtained by calculating weights, and then sentiment classification was performed. Wu et al. [7] help to perform sentiment analysis of social media content by using Web resources to build an easily maintained and expandable sentiment dictionary.

In the study of sentiment analysis based on machine learning, because deep learning has the advantages of local feature abstraction and memory function, it can avoid a large number of artificial feature extraction and other advantages. Research has applied deep neural network-based text classification methods to sentiment analysis, the most popular of which are CNN and RNN models. Hu et al. [1] used a recurrent neural network to perform multimodal fusion sentiment analysis on English text. Kim et al. [8] proposed an improved CNN for text classification. Zhao et al. [9] improved the proposed cyclic random walk network model by using back propagation method, and successfully classified sentiment on TWitter's blog posts, showing good performance. Bai et al. [10] used the Bilstm recursive neural network to extract features and combined context semantics to perform position detection on Sina Weibo Chinese text sentiment. Hu et al. [11] combined a long-term and short-term memory network in a recurrent neural network with a feedforward attention model and proposed a text sentiment classification scheme. Chen et al. [12] used the proposed multi-channel convolutional neural network model to classify sentiment in Chinese microblog text.

CNN, RNN and other neural networks have achieved good results in the field of sentiment analysis in natural language processing, but there are still some problems. For example, CNN was originally widely used in the field of image processing, and it is not very suitable for processing text sequence problems. Although deep neural networks RNNs can achieve good results in processing text classification tasks, RNNs have long-term dependencies and may face gradient explosion or gradient disappearance problems [13]. In order to solve this series of problems, many neural network variants such as LSTM and GRU have been proposed successively, and have been successfully applied in the field of sentiment analysis. LSTM is able to memorize the context of a sequence, which has obvious advantages for the emotional feature extraction of text sequences, and can solve the problem of vanishing gradients. As a variant of LSTM, GRU has a simpler structure than LSTM, with fewer parameters and faster convergence, and can solve the problem of long-distance dependence. In the latest text classification research, it is found that compared with GRU, the smallest gated unit MGU has the advantages of simpler structure, fewer parameters, and less training time. It is very suitable for tasks with strong time dependence, while using fewer parameters. The MGU model can reduce the workload of selecting parameters and improve the generalization ability of the model [14].

The MGU model does not fully learn the sequence-related features such as the sequence in which the text is generated in time. The convolutional neural network CNN can extract the features of the data through convolution operations, and can enhance certain features of the original data and reduce the noise. Impact. When a similar human brain recognizes a picture, it does not recognize the whole picture at once, but it first perceives each feature in the picture first, and then integrates the parts at a higher level. Operation to get global information.

The Attention mechanism is a resource allocation mechanism, that is, at a certain time, your attention is always focused on a fixed position on the screen without paying attention to other parts. The Attention mechanism was originally only applied to the task of image recognition in computer vision, and then applied to image-text conversion. In natural language processing, Attention is often used in combination with various neural networks to help neural networks focus on some important information in text sequences.

Therefore, in this paper, a new attention C_MGU model is proposed by combining the network structure of CNN and the minimum gating unit MGU and introducing the attention mechanism. This model combines the convolutional layer and the smallest gated unit MGU in CNN with a unified architecture, and adds the attention mechanism between the convolutional layer and the MGU layer. The new model not only reflects the advantages of the CNN model and the smallest gated unit MGU, but also uses the attention mechanism to further optimize the semantic feature representation. The experimental results show that the new model can not only highlight the key information of the text, but also mine richer semantics, and has a better performance in the sentiment classification of the text.

2   Recurrent Neural Network with Gate Structure

2.1   LSTM and GRU

RNN has been proven to be very successful in the field of natural language processing in practice, but when faced with long sequences of text, the gradient of the hidden layer variables of RNN may attenuate or explode. Although gradient clipping can cope with gradient explosion, it cannot solve gradient attenuation. Therefore, given a text sequence, it is actually more difficult for a recurrent neural network RNN to capture text elements (words or words) with a large distance between two moments. LSTM (Long Short Term Memory) memory unit is based on the RNN memory unit with a gate control mechanism. Its structure is shown in Fig. 1. It implements three gate calculations, namely the forget gate, input gate and output gate [15]. The Forget Gate is responsible for deciding how many unit states from the previous moment to the unit state for the current moment; the Input Gate is responsible for deciding how many unit states from the current moment to the unit state at the current moment; Output Gate Responsible for determining how many outputs the unit status has at the current moment. Each LSTM contains three inputs, namely the unit status at the last moment, the output of the LSTM at the moment in time, and the input at the current moment. The historical information can be filtered to solve the problem of gradient disappearance.

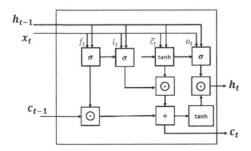

Fig. 1. LSTM

Because there are more learning parameter settings in LSTM and longer training time, GRU is proposed as an improved version of LSTM [16]. Compared with LSTM, the GRU has one less gated unit and fewer parameters, so the computing time and convergence speed are greatly improved. Its structure is shown in Fig. 2.

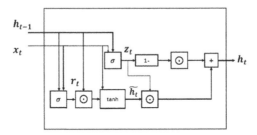

Fig. 2. GRU

The GRU model is shown in the figure. It has only two gates, namely the update gate and the reset gate, which are z and r in the figure. The update gate is used to control the degree to which the state information of the previous moment is brought into the current state. The larger the value of the update gate is, the more state information is brought into the previous moment. The reset gate is used to control the degree of ignoring the state information of the previous moment. The smaller the value of the reset gate is, the more it is ignored. At time t, for a given input x_t, the hidden output h_t of the GRU is calculated as follows:

$$r_t = \sigma \left(W_r \cdot \left[h_{t-1}, x_t \right] \right) \tag{1}$$

$$z_t = \sigma\left(W_z \cdot \left[h_{t-1}, x_t\right]\right) \tag{2}$$

$$\tilde{h}_t = \tanh\left(W_{\tilde{h}} \cdot \left[r_t * h_{t-1}, x_t\right]\right) \tag{3}$$

$$h_t = (1 - Z_t) * h_{t-1} + Z_t * \tilde{h}_t \tag{4}$$

Where [] indicates that the two vectors are connected, W is the weight matrix of the connection layer, and tanh is the activation function.

2.2 Minimum Gated Unit MGU

The minimum gated unit MGU is a simplified recurrent neural network structural unit. The gate number is the smallest in any gated unit, and there is only one forget gate. The input gate is merged into the forget gate. The structure is shown in Fig. 3.

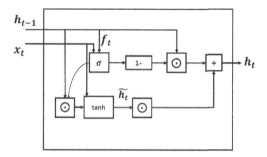

Fig. 3. MGU

As can be seen from the structure diagram of the MGU, the structure of the MGU is obviously simpler than that of the LSTM and GRU. The LSTM has four sets of parameters to be determined, and the GRU needs three sets of parameters, but the MGU only needs two sets of parameters. In the gating unit, the recurrent neural network has a simpler structure, fewer parameters, and faster training.

3 Sentiment Analysis Model Based on Attention C_MGU

3.1 Model Design

Combining the respective advantages of CNN and MGU, in order to achieve the goal of text sentiment analysis, this paper proposes a network structure that combines CNN with the smallest gating unit MGU and introduces the attention mechanism. The model structure is shown in Fig. 4. First, the model captures the preliminary feature representation of the extracted text through the CNN's convolutional layer module, and then uses the Attention mechanism and the minimum gating unit MGU module to strengthen the key information of the preliminary feature representation of the text. With further optimization, the final deep feature representation of the text is generated in the hidden layer of the MGU, and it is input to the Softmax layer for regression processing, and the classification of the text is finally completed.

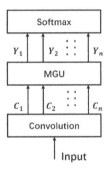

Fig. 4. C_MGU attention

3.2 Algorithm Design

The specific process of the model structure proposed in this paper is: firstly capture the preliminary feature representation of the text through the CNN's convolution layer, and then use the attention mechanism and the MGU model to obtain the deep features of the text with keyword discrimination in the hidden layer of the MGU Indicates that the state of the hidden layer is finally input to the Softmax layer for regression classification processing to complete the classification processing of the text.

Convolution Layer Feature Extraction. The first module of this model is to extract the preliminary feature representation of the text through the convolutional layer in CNN. By defining the region vector RSV to maintain the original sequence corresponding to the input sentence, the MGU model of the specified sequence input provides a reasonable input. The purpose of using the convolution layer in this module is to ensure a reasonable sequence vector as the input vector for subsequent MGUs. The pooling layer will destroy the original word order of the original sentence and cannot be used as a reasonable input for the MGU, so only the convolution layer is used. Define filters in a one-dimensional convolutional layer to extract local feature representations of different text positions, $w_j \in R^D$ means D-dimensional word embedding representing the jth word in a sentence. $x \in R^{I \times D}$ represents an input sentence of length L, $f \in R^{K \times D}$ is the filter with length K in the convolutional layer. The region sequence vector is marked with S_i and is composed of K word embeddings starting from the position i of the input sentence.

$$s_i = \left[w_i, w_{i+1} \cdots, w_{i+k-1} \right] \tag{5}$$

The input sentence is processed by the filter of the convolution layer to generate a new feature map $c \in R^{L-K+1}$. The conversion formula is as follows:

$$c_i = ReLU \left(s_i^\circ f + \theta \right) \tag{6}$$

Where θ represents the offset and the ReLU function is a non-linear activation function. Use N filters of the same length to generate the feature matrix:

$$F = \begin{bmatrix} c_{11} & \cdots & c_{1i} \\ \cdots & \cdots & \cdots \\ c_{n1} & \cdots & c_{ni} \end{bmatrix} \tag{7}$$

Attention Layer. With the development of deep learning in recent years, the attention mechanism has been widely used in image caption generation, machine translation, speech recognition and other fields, and has achieved outstanding achievements [17]. The Attention mechanism simulates the attention model of the human brain. For example, when we admire the painting, we can see the whole picture, but when we look closely, the glasses focus on only a small part. The main concern lies on this small pattern, that is, when there is a certain weight differentiation, the human brain's attention to the entire painting is not balanced. The Attention mechanism is a mechanism that highlights local important information by assigning sufficient attention to key information [18]. In the model proposed in this paper, each output element is obtained by clicking the formula:

$$y_i = F(C_i, y_1, y_2 \cdots y_{i-1}) \tag{8}$$

$$C_i = \sum_{j-1}^{T} a_{ij} S(x_j) \tag{9}$$

$$a_{ij} = \frac{\exp(e_{ij})}{\sum_{k-1}^{T} \exp(e_{ik})} \tag{10}$$

$$e_{ij} = score(s_{i-1}, h_j) = vtanh(Wh_j + Us_{i-1} + b) \tag{11}$$

Among them, C_i is an input sentence $x_1, x_2 \ldots$ which is obtained by non-linear transformation after the convolution layer operation, represents the corresponding hidden state of the j word embedding, T represents the number of input sequence elements, and a_{ij} represents the input j corresponding to the output Y_i attention distribution probability. Among them, e_{ij} is a verification model, which aims to reflect the influence evaluation score of the j input on the i output, h_j is the hidden state of the j input in the convolution layer, W and U are weight matrices, and b is Offset. The semantic coding formed by the Attention mechanism will be used as the input of the MGU module.

MGU Layer. After the Attention mechanism, this paper uses a simplified recurrent neural network structural unit, namely the smallest gate unit MGU. Compared with LSTM and GRU, MGU has fewer parameters [19]. At time t, the MGU model calculates the current state as:

$$h_t = (1 - f_t) \odot c_{t-1} + f_t \odot \tilde{h}_t \tag{12}$$

The forget gate controls the degree of memory forgetting at the last moment and how much new information is added. The forget gate is expressed as:

$$f_t = \sigma(W_f + U_f h_{t-1} + b_f) \tag{13}$$

$$\tilde{h}_t = tanh(W_h x_t + f_t \odot (U_t h_{t-1}) + b_h) \tag{14}$$

Compared with LSTM and GRU, MGU has fewer parameters, simpler structure, and faster convergence. Through the above model, a deep-level feature representation of the text can be obtained. Finally, Softmax is used to perform regression to obtain the final text classification result.

4 Experimental Analysis

4.1 Data Set

For the sentiment classification task, the publicly used IMDB and Sentiment140 datasets are selected in this paper. The IMDB film review dataset is a binary sentiment classification dataset, with positive and negative reviews each accounting for 50%. The Sentiment140 dataset is also a dataset that can be used for sentiment analysis. It is composed of 160,000 tweets, and emoticons have been removed. In this paper, 20,000 pieces of data are selected according to the principle of positive and negative balance, and divided into a training set and a test set according to a ratio of 8:2.

4.2 Experimental Settings

This article uses the above two data sets for pre-training. After word segmentation and processing of the above corpus, the word vector is trained using Word2vec. The dimension of the word vector is 100, and the vector representation of the text word can be obtained. If a word that is not in the word vector vocabulary appears when the model is running, random initialization is used. Since the convolutional layer module requires fixed-length input, this article defines a Maxsize to indicate the maximum length of the allowed text sentence. Through the threshold Maxsize to limit the text length, a fixed-length input text sentence can be obtained. In addition, this article Set the region sequence length to 5 and the number of filters to 2.

For the C_MGU model based on the attention mechanism, this paper defines the context vector in the attention mechanism as 100 dimensions, the number of MGU hidden units is 100, the learning rate is 0.01, and the batch-size is 20. The hidden layer state of the MGU is used as the input of the Softmax. The interval steps of the classifier are set to 100 and the batch-size is set to 20. The stochastic gradient descent method is used as the optimization method.

This experiment uses the Keras deep learning framework. The bottom layer is TensorFlow. The TensorFlow platform integrates CNN, RNN and LSTM, GRU, MGU and other deep learning models. It is implemented using Python programming.

4.3 Analysis of Experimental Results

Parameter Setting Comparison Experiment. The word vector is a more important text processing task. Its dimension directly affects the classification accuracy of the model. In this paper, Word2Vec is used to train the word vector. The text is given an initial value so that the network can better learn and adjust parameters during the training process. To make the model have better classification performance. Therefore, the choice of the word vector dimension is particularly important. The Att-C_MGU and CNN, LSTM, GRU and other models proposed in this paper perform on the data set IMBD when the word vector dimensions are different, thereby determining the choice of the word vector dimension in this paper problem. The specific impact of different word vector dimensions on the model's classification effect on the IMBD dataset is shown in Fig. 5.

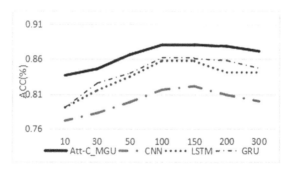

Fig. 5. Influence of word vector dimensions

It can be seen from Fig. 5 that when Word2Vec is used to train word vectors, when the word vector dimension is less than 100, the accuracy of the selected model has a significant upward trend. When the word vector dimension is greater than 100, the selected model is in the data The classification performance of the set shows fluctuations, and some rises more gently, but as the dimension of the word vector increases, the model does not learn the feature information of the word vector well, but it decreases, so this article The dimension of the word vector is selected as 100.

After determining the size of the word vector, the choice of model iterations in this article is also explained through experiments. The number of iterations is the number of times the entire training set is trained. As the number of iterations increases, the results of the network model gradually approach the optimal, When the number of iterations exceeds a certain number of times, it will lead to overfitting and reduce the generalization ability of the model.

It can be seen from Fig. 6 that with the increase of the number of iterations, the accuracy of the model will gradually increase. When the number of iterations is about 30, the accuracy will stabilize, the change will be relatively gentle, and the performance effect will be better.

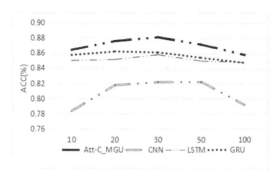

Fig. 6. Iteration times

172   D. Wang et al.

Attention Mechanism Affects Experiments. In the experiments in this paper, after pre-processing the text data, jieba word segmentation is used to process the text sentence, then Word2vec is used to train the word vector, and the trained word vector is used as the model input. After the attention C_MGU model proposed in this paper, Finally, the output results are compared with the labeled data to show the effectiveness of the model. In order to illustrate the necessity of the attention mechanism and the role of the attention mechanism, the performance of the attention C_MGU model and the C_MGU without the attention mechanism were compared on the IMBD and Sentiment140 datasets. The experimental results are shown in Fig. 7. As shown. As can be seen from Fig. 7, due to the introduction of the attention mechanism in the sentiment analysis algorithm, the performance of the algorithm has been improved to a certain extent, and the accuracy of the IMBD, IMBD2, and Sentiment140 dataset classification has been improved by 2.8%, 2.3%, 1.7%. This illustrates the necessity of the Attention mechanism for the new model.

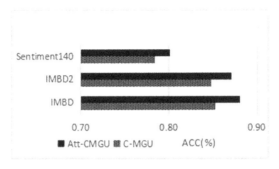

Fig. 7. Impact of attention mechanism

5   Conclusions

This paper proposes a C_MGU-based hybrid neural network model. By performing sentiment analysis experiments on IMBD film review data and Sentiment140 data, it compares accuracy and F1 values with other mainstream text classification methods. Effective. Compared with GRU and LSTM, MGU has certain advantages in the field of text sentiment analysis due to its advantages such as small calculation amount and fast convergence speed.

In the next research work, we will carry out research in the following aspects: In the research of this paper, there are only two types of emotional polarity: positive and negative, the classification of emotion is not refined, and the issue of emotional intensity is not considered; During data processing, some data with only emojis or only pictures and no text descriptions were discarded, which could not be comprehensively analyzed from multiple angles, which caused some limitations to the model. The next step will be in-depth research in sentiment analysis of multimodal text.

References

1. Hu, A., Flaxman, S.: Multimodal sentiment analysis to explore the structure of emotions. In: Proceedings of the 24th ACM SIGKDD International Conference on Knowledge Discovery & Data Mining, KDD 2018, London, United Kingdom, 19–23 August 2018, pp. 350-358. ACM Press (2018)
2. Chen, L., Guan, Z.Y., He, J.H., et al.: Research progress of sentiment classification. Comput. Res. Dev. **54**(6), 1150–1170 (2017)
3. Chen, L., Guan, Z.Y., He, J.H., et al.: A survey on sentiment classification. J. Comput. Res. Dev. **54**(6), 1150–1170 (2017)
4. Li, S., Huang, C.R., Zhou, G.: Employing personal impersonal views in supervised and semisupervised sentiment classification. In: Proceedings of the 48th Annual Meeting of the Association for Computational Linguistics, Uppsala, 11–16 July 2010, pp. 414–423. ACL, Stroudsburg (2010)
5. Zhang, Y., Fu, J., She, D., Zhang, Y., Wang, S., Yang, J.: Text emotion distribution learning via multi-task convolutional neural network. In: IJCAI (2018)
6. Zhang, S., Wei, Z., Wang, Y., et al.: Sentiment analysis of Chinese micro-blog text based on extended sentiment dictionary. Future Gener. Comput. Syst. **81**, 395–403 (2017). S0167739X17307835
7. Wu, L., Morstatter, F., Liu, H.: SlangSD: building and using a sentiment dictionary of slang words for short-text sentiment classification. Lang. Resour. Eval. **52**(6), 1–14 (2016)
8. Kim, Y.: Convolutional neural networks for sentence classification. Eprint Arxiv arxiv.org/abs/1408.5882 (2014)
9. Zhuang, Y., Zhao, Z., Lu, H., et al.: Microblog sentiment classification via recurrent random walk network learning. In: Twenty-Sixth International Joint Conference on Artificial Intelligence (2017)
10. Bai, J., Li, Y., Ji, D.H.: Attention-based BiLSTM-CNN Chinese Weibo position detection model. J. Comput. Appl. Softw. **35**(3), 266–274 (2018)
11. Hu, R.L., Rui, L., Qi, X., et al.: Text sentiment analysis based on recurrent neural network and attention model. Appl. Res. Comput. (11) (2019)
12. Chen, K., Liang, B., Ke, W.D.L., et al.: Sentiment analysis of Chinese Weibo based on multi-channel convolutional neural network. J. Comput. Res. Dev. **55**(05), 55–67 (2018)
13. Mikolov, T., Sutskever, I., Chen, K., et al.: Distributed representations of words and phrases and their compositionality. In: Proceedings of the 27th Annual Conference on Neural Information Processing Systems. Cambridge, Nevada, 5–10 December 2013, pp. 3111–3119. MIT Press, Cambridge (2013)
14. Cleeremans, A., Servan-Schreiber, D., Mcclelland, J.L.: Finite state automata and simple recurrent networks. Neural Comput. **1**(3), 372–381 (1989)
15. Hochreiter, S., Schmidhuber, J.: Long short-term memory. Neural Comput. **9**(8), 1735–1780 (1997)
16. Cho, K., Merrienboer, B.V., Gulcehre, C., et al.: Learning phrase representations using RNN encoder-decoder for statistical machine translation. In: Proceedings of the Empirical Methods in Natural Language Processing, Doha, 25–29 October 2014, pp. 1724–1735. ACL, Stroudsburg (2014)
17. Wang, C., Jiang, F., Yang, H.: A hybrid framework for text modeling with convolutional RNN. In: The 23rd ACM SIGKDD International Conference. ACM (2017)
18. Seo, S., Huang, J., Yang, H., et al.: Interpretable convolutional neural networks with dual local and global attention for review rating prediction. In: The Eleventh ACM Conference. ACM (2017)
19. Zhou, G.B., Wu, J., Zhang, C.L., et al.: Minimal gated unit for recurrent neural networks. Int. J. Autom. Comput. **13**(3), 226–234 (2016)

Inception Model of Convolutional Auto-encoder for Image Denoising

Diangang Wang[1], Wei Gan[1], Chenyang Yan[2(✉)], Kun Huang[1], and Hongyi Wu[3]

[1] State Grid Sichuan Information and Communication Company, Shanghai 610041, China
[2] School of Computer Science and Technology, Shanghai University of Electric Power, Shanghai 200090, China
857322130@qq.com
[3] School of Information and Electronic Engineering, Zhejiang University of Science and Technology, Zhejiang 310023, China

Abstract. In order to remove the Gaussian noise in the image more effectively, a convolutional auto-encoder image denoising model combined with the perception module is proposed. The model takes the whole image as input and output, uses the concept module to denoise the input noise image, uses the improved concept deconvolution module to restore the denoised image, and improves the denoising ability of the model. At the same time, the batch normalization (BN) layer and the random deactivation layer (Dropout) are introduced into the model to effectively solve the model over fitting problem, and the ReLu function is introduced to avoid the model gradient disappearing and accelerate the network training. The experimental results show that the improved convolution neural network model has higher peak signal-to-noise ratio and structure similarity, better denoising ability, better visual effect and better robustness than the deep convolution neural network model.

Keywords: Convolutional auto-encoder · Inception module · Image denoising · Peak signal to noise ratio · Structural similarity

1 Introduction

With the rapid development of computer technology and Internet technology, people's daily life is full of all kinds of information. According to the investigation and research, among all the external information obtained by human beings, vision system accounts for more than 70% [1], so the acquisition, processing and use of image information is particularly important. Image denoising is an important research topic in the field of image processing. While removing the noise, we try to keep the important information in the image. Digital image processing can be generally divided into space-based processing and transform based processing [2]. The denoising method based on the spatial domain is to operate on the gray space of the original image, and process the gray value of the pixel directly. The common methods include mean filter, median filter and image

© ICST Institute for Computer Sciences, Social Informatics and Telecommunications Engineering 2020
Published by Springer Nature Switzerland AG 2020. All Rights Reserved
J. Liu et al. (Eds.): MobiCASE 2020, LNICST 341, pp. 174–186, 2020.
https://doi.org/10.1007/978-3-030-64214-3_12

denoising based on partial differential. Median filter can effectively filter salt and pepper noise and mean filter is suitable for filtering Gaussian noise. The denoising method based on transform domain is to transform the source image first, such as Fourier transform, wavelet transform, etc. Subsequent paragraphs, however, are indented.

At present, many image denoising methods have been proposed by scholars at home and abroad. At present, the BM3D (block matching and 3D) [3] algorithm with better denoising effect is to divide the image into blocks of certain size, merge the blocks with similar characteristics into three-dimensional arrays, process the three-dimensional arrays by three-dimensional filtering method, and obtain the denoised image by inverse transformation; Schuler [4] and others put forward MLP (multilayer perceptron) model, which uses image preprocessing and multilayer perceptron Through the combination of network learning model. The algorithm proposed by Burger [2] uses MLP in image denoising. Chen et al. [5] Proposed TNRD (traditional nonlinear reaction diffusion) model, expanded sparse coding and iterative methods into forward feedback network, and achieved good image denoising effect.

In recent years, research shows that as a typical representative of deep learning, Auto-encoder (AE) is mainly used to learn the compression and distributed feature expression of given data through unsupervised learning, so as to reconstruct the input data [6]. Based on the auto-encoder, researchers have derived a variety of auto-encoders. Hinton [7] and others improved the original shallow structure, proposed the Inception and training strategy of deep learning neural network, and then produced the Denoising Auto-Encoder (DAE); in 2007, Benjio [8] proposed the Inception of Sparse Auto-Encoder (SAE); in addition, there were Marginalized Denoising Auto-Encoder (MDA) and Stacked Sparse Denoising Auto-Encoder (SSDA) [9].

In this paper, a neural network denoising model based on a convolutional autoencoder is used to speed up the operation speed of the network. This network model has changes in the size of the convolution layer. The Inception module is used to expand the network width to better extract noise image features. The network structure of the improved Inception deconvolution module is used. Batch normalization and random inactivation are used to prevent overfitting. The length and width of the convolution layer are inversely proportional to the number of feature maps, which greatly reduces the number of network parameters. The data set uses VOC2012. Due to the huge content of the data set, 1000 pictures are randomly selected as the training set, 700 of which are used as the training set, and 300 pictures are used as the test set. Classical image data is used for comparative experiments. It is proved by experiments that the algorithm structure in this paper is more robust to denoising and has better denoising effect.

2 Network Structure of Convolutional Auto-encoder

Image denoising is the process of processing and restoring the noisy image. In this paper, the lightweight network structure is used to achieve excellent denoising effect and the deep learning network structure of four-layer convolutional auto-encoder is used. In order to speed up network training, the data in the data set is divided into 20×20 sizes. After adding the noise, the original image content is stored in different H5 files to speed up file reading and complete network training better.

2.1 Generate Noise Image

Gaussian noise is a kind of random noise which accords with normal distribution, and it is also the most common noise distribution. As shown in formula (1),

$$
\begin{aligned}
Z &\sim N(\mu, \sigma) \\
T_{(h,w,c)} &= X_{(h,w,c)} + k \cdot Z \\
X_{(h,w,c)} &= \begin{cases} 0, & T_{(h,w,c)} < 0 \\ 255, & T_{(h,w,c)} > 255 \\ T_{(h,w,c)}, & others \end{cases}
\end{aligned}
\tag{1}
$$

Among them, Z is the noise data, which conforms to the normal distribution with the expectation of μ and the variance of σ. k is the noise intensity, and $X_{(h,w,c)}$ is the image pixel. Finally, the value of image pixel after noise is added into the formula to limit, so as to avoid data overflow [10]. In this paper, we use the Gaussian noise data set with noise level of 25 to denoise the data. As shown in Fig. 1, we can see the difference between the noisy image and the original image. In this paper, we remove the image noise based on the noise level of 25.

Fig. 1. Comparison between the original image and the noise image

2.2 Multi Feature Extraction Inception Module

Inspired by popular image processing algorithms such as VGG Net and GoogLeNet, the InceptionV3 module is used to extract image features and restore images, and good results are achieved. In order to solve the problem of increasing the depth and width of the network while reducing the parameters, the Inception module mainly improves the traditional convolution layer in the network, the structure diagram is shown in Fig. 2. The inception module calculates several different transformation results on the same input map in parallel, and connects their results into one output. Using the inception module is conducive to extract as much feature information as possible from different convolution kernel sizes of noisy images, and provides better generalization ability for the model network. Therefore, this paper improves on the basis of inception, changes the original convolution layer to deconvolution layer for up sampling operation, uses the combination of small convolution kernels of $1 \times 1, 3 \times 3, 5 \times 5$ to reduce the channel dimension of the feature image, better restores the feature image, and makes it closer to the original image.

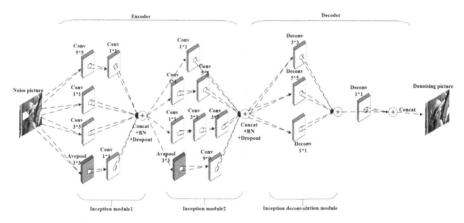

Fig. 2. Network structure of convolutional auto-encoder denoising

Different from the up sampling operation, this paper uses two-layer inception module to process noise image. This brings another problem: the number of feature maps in each layer increases, and the cost of computation increases greatly. Therefore, this paper makes the following settings for the inception module:

1) Each convolution layer of Inception is added to the ReLu activation function, which simplifies the calculation process. The dispersion of activity makes the calculation cost of Inception module decrease;
2) Add batch normalization (BN) and random deactivation layer (Dropout). BN layer can accelerate the training speed of inception network many times, improve the generalization ability of the network, normalize the output to the normal distribution of n (0, 1), reduce the distribution of internal neurons, accelerate the training speed, and produce more stable nonlinear output. In the experiment, it is found that the training PSNR is not stable when only BN layer operates, and the problem of non data set and non verification set is considered After that, the Dropout layer is used to solve the over fitting phenomenon in the model training process. The results show that the Dropout layer can reduce the PSNR instability. In Dropout learning process, part of the weight or output of the hidden layer is randomly zeroed to reduce the interdependence between nodes.

2.3   Design of Convolutional Auto-Encoder Network Based on Inception Module

In order to make the denoising network model be able to process natural images, the data of each image is transformed into a three-dimensional matrix. The convolutional auto-encoder is divided into two parts: decoder and encoder. There are four layers in total. The structure of the convolutional auto-encoder denoising network based on the Inception module is shown in Fig. 2. The advantage of the network is that it uses auto-encoder structure, encoder and decoder of coding layer, the first two layers of InceptionV3 classic structure in Encoder and deconvolution module in Decoder. The advantage of this module is that it can restore the noise image features extracted from the encoder to a greater extent, and it can restore the original image features better than one layer deconvolution. The specific network settings are as follows:

Encoder:

The first layer: it consists of five different scale convolution layers and an average pooling layer to form the Inception module. It can enlarge the width of convolution auto-encoder, extract information of different sizes of image using multiple convolution kernels, and fuse them to get better representation of image. The first layer of convolution layer is $5 \times 5 \times 32$, $1 \times 1 \times 64$, and the image output channel is 64; the second layer is $3 \times 3 \times 64$, and the image output channel is 64; the third layer is $1 \times 1 \times 64$, and the image output channel is 64; the fourth layer is the average pooling layer, with a step size of 1, and the pooling layer is 3×3, followed by a layer of $1 \times 1 \times 32$ convolution layer, and the image output channel is 32. Input of each layer is added with standardization, and padding is same using ReLu function to prevent the gradient from disappearing. Finally, the Concat layer is used to connect, the standardized BN layer is added, and Dropout is used to prevent over fitting. At this time, the output channel of the picture is $64 + 64 + 64 + 32 = 224$;

The second layer: using the structure of the second module in Inception V3, the first layer of Conv is $1 \times 1 \times 64$, the second layer is $1 \times 1 \times 48$, $5 \times 5 \times 64$; the third layer is $1 \times 1 \times 64$, $3 \times 3 \times 96$, $3 \times 3 \times 96$, the fourth layer is the average pooling layer, the pooling layer size is 3×3, and the step size is 1. After the pooling layer is connected with a convolution layer, the convolution core size is 1×1, and the channel size is 32. At last, Concat layer is used to connect, standard BN layer is added, and Dropout is used to prevent over fitting. After the Inception module of this layer, the output picture channel is $64 + 64 + 96 + 32 = 256$.

Decoder:

The first layer: the upper sampling layer is implemented by the improved Inception module using deconvolution. It is composed of four different dimensions of anti convolution layers, which are $3 \times 3 \times 16$, $5 \times 5 \times 16$, $1 \times 1 \times 16$, step size is set to 2, and Concat layer is used for connection. Using the improved Inception module for deconvolution can make the feature fusion better. At this time, the shape of the picture is $20 \times 20 \times 64$, adding BN layer for standardization operation;

The second layer: use deconvolution to realize the upper sampling layer, and use the upper sampling layer to process the image of decoder. In order to get the same size of the original image, use the upper sampling layer of the first layer to realize, and restore the image to the original size. At this time, the image shape is $20 \times 20 \times 1$.

In conclusion, in order to improve the robustness of image denoising, the convolution operation of the Inception module is introduced to improve the convolution operation in the Inception module, better feature extraction of noise image, use the ReLu function to prevent the gradient from disappearing, introduce BN and Dropout operation to prevent network over fitting, improve the overall denoising performance of the model, and shorten the training time.

The flow of image denoising using the network is shown in Fig. 3. With the increase of training times, the verification set is used to evaluate whether the model is over fitted. The specific operation is: set the number of nodes to 500, after training the corresponding parameters through the training set, the verification set is used to detect the error of the model, and then change the number of nodes. If the error of the model is greater than 100% or less than 0%, stop the network immediately and make corresponding modification.

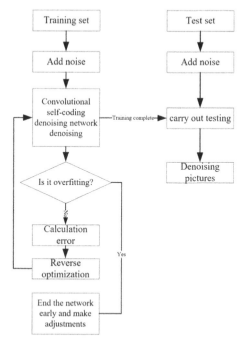

Fig. 3. Training flow of decontamination network of convolutional auto-encoder

3 Experiments and Results

3.1 Experimental Data

In the image denoising experiment based on convolutional auto-encoder, VOC2012 data set is very large, so 1000 of them are randomly selected as data sets, 700 of them are training sets and 300 are test sets. At the same time, 10 standard images commonly used in the field of image denoising are used as reference images for comparative experiments. All the images in voc2012 are color images, while the gray image is used in this paper. Therefore, it is necessary to convert the color image into gray image and add Gaussian noise with noise level of 25. In order to facilitate training, the input image is cut into 20×20 sub image blocks, and the cut-out image is stored in an H5 file every five original images, which is convenient for model reading and training.

3.2 Experimental Environment and Evaluation Criteria

The experimental environment system is configured as windows 10 system, the processor is Intel Core i7-3370 CPU, and the memory is 8 GB.

Peak Signal to Noise Ratio (PSNR) and Structural Similarity (SSIM) are used as noise reduction evaluation indexes. As shown in formula (3) and (4).

1) Mean Square Error (MSE)

$$MSE = \frac{1}{M \times N} \sum_{i=1}^{M} \sum_{j=1}^{N} \left(\hat{f}(i,j) - f(i,j) \right)^2 \qquad (2)$$

MSE represents the mean square error of the current image $\hat{f}(i,j)$ and the reference image $f(i,j)$. M and N are the height and width of the image respectively.

2) Peak Signal to Noise Ratio (PSNR)

$$PSNR = 10 \log_{10} \left(\frac{(2^n - 1)^2}{MSE} \right) \qquad (3)$$

Where MSE (formula (2)) is the mean square error between the original image and the denoised image, n is the number of bits per pixel, generally 8, that is, the number of pixel gray scale is 256. The unit is dB, the larger the PSNR value is, the less the representative distortion is.

3) Structural Similarity (SSIM)

$$SSIM(x, y) = \frac{(2\mu_x + C_1)(2\sigma_{xy} + C_2)}{\left(\mu_x^2 + \mu_y^2 + C_1 \right) \left(\sigma_x^2 + \sigma_y^2 + C_2 \right)} \qquad (4)$$

Where μ_x is the average of x, μ_y is the average of y, σ_x^2 is the variance of x, σ_y^2 is the variance of y, σ_{xy} is the covariance of x and y. $c_1 = (k_1 L)^2$ $c_2 = (k_2 L)^2$ is a constant used to maintain stability, L is the dynamic range of pixel value, $k_1 = 0.01, k_2 = 0.03$. The range of similar structure is 0–1. The larger the value of similar structure is, the closer the two images are.

3.3 Influence of Network Model on Denoising Performance

The Influence of Inception Module on Noise Reduction Performance. In this paper, the network uses the Inception module to extract the features of images and in order to show the feature extraction ability of multiple Inception structures in this paper, the common convolutional auto-encoder, one layer of Inception module and multiple Inception modules in this paper are used for PSNR comparison. The experimental results are shown in Fig. 4. In the contrast experiment, the same decoding layer is set up, which is 3 × 3 × 32 and 3 × 3 × 1 anti convolution layer respectively. Different methods are used in the encoder. The encoder adopts the structure of two layers of convolution layer, and the general auto-encoder uses two 3 × 3 convolution layers as the encoder; the first layer of Inception module uses the first Inception module and the second layer uses the 3 × 3 convolution layer; the two layers of Inception module uses the module used in this paper. Using the same experimental environment and training set, the training process outputs the training PSNR. After 500 times of training, it can be seen from Fig. 4 that the PSNR value of the algorithm in this paper keeps rising during the training process, up to 25 dB or more, while the initial stage of the ordinary convolutional auto-encoder is poor, and it slowly drops in the early stage of the sudden rise and the later stage, and finally it is about 19 dB; the first layer of Inception rises slowly after the fluctuation in the early stage, and finally it is about 21 dB. The two-layer Inception module used in this paper is superior to the other two methods from the beginning, and finally it is gentle at about 23 dB. From the experimental results, we can see that the more training times, the better stability and robustness of this algorithm, so it shows better denoising results.

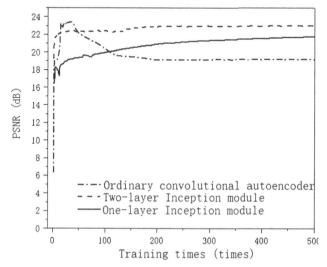

Fig. 4. Comparison of the effects of different coding layers on the model

The Influence of the Deconvolution Module of Inception on the Performance of Denoising. In this paper, we use the Inception deconvolution to feedback the extracted features. Compare the influence of one layer deconvolution and Inception deconvolution module on image denoising. The design of the coding layer of the comparison experiment is consistent with the encoder of this paper. The decoding layer uses a deconvolution layer with a convolution core of 3×3 and the deconvolution module of this paper to carry out the comparison experiment. The experimental results are shown in Fig. 5. From Fig. 5, it can be seen that the ordinary one-layer deconvolution in the early training is relatively stable, while the Inception deconvolution module fluctuates briefly. In the later training, the effect of using the Inception module is better than that of the ordinary one-layer deconvolution, and the final stability is about 24 dB. The overall effect is very good when using the Inception deconvolution module.

3.4   Comparative Experimental Analysis

In order to verify the robustness of the method in this paper, 10 classic test images are selected for simulation experiment, and compared with literature [11], literature [12] and

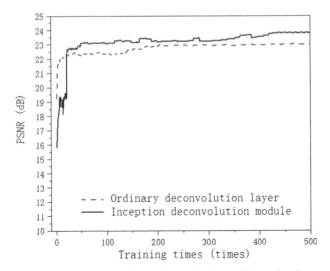

Fig. 5. Comparison between one layer deconvolution and Inception deconvolution

literature [13], as shown in Table 1 and Table 2, Table 1 is the peak signal-to-noise ratio of ten images, and Table 2 is the structural similarity of each method. Both literature [11] and literature [12] use deep convolution neural network for image denoising, as shown in Table 1 and Table 2, the algorithm in this paper shows good denoising effect, with an average increase of PSNR by 11.088 and SSIM by 0.451 compared with the original image; literature [11] and literature [12] use 5-layer deep convolution neural network. The difference is that the first three layers of literature [11] are convolutions, the last two layers are anti convolutions, and literature [12] uses five convolutions to denoise. Compared with literature [11] and literature [12], PSNR and SSIM increased by 2.813 and 0.821, respectively, and 5% and 1.9% respectively. In reference [13], the PSNR and SSIM were increased by 2.626% and 1.1% respectively under the same experimental environment by using one-layer Inception module and five-layer convolution layer.

Table 1. Peak signal to noise ratio (PSNR) of each method for ten images

PSNR	House1	Woman1	Woman2	Man	Camera	Lena	Barbara	Boat	House2	Girl
Original drawing	20.18	20.21	20.34	20.22	20.58	20.23	20.31	20.30	20.35	20.63
Document [11]	28.13	28.36	30.38	28.01	27.99	29.19	26.58	27.68	30.08	29.7
Document [12]	29.19	29.32	30.85	30.11	30.28	30.21	29.58	30.22	30.77	30.09
Document [13]	29.46	29.15	29.22	28.84	28.15	28.86	29.03	28.26	28.69	28.31
This paper	30.12	31.55	32.08	31.69	30.59	32.54	30.04	32.13	31.82	31.67

Table 2. SSIM of ten images

SSIM	House1	Woman1	Woman2	Man	Camera	Lena	Barbara	Boat	House2	Girl
Original drawing	0.43	0.49	0.42	0.53	0.49	0.49	0.57	0.53	0.42	0.47
Document [11]	0.90	0.89	0.92	0.87	0.89	0.91	0.85	0.87	0.85	0.90
Document [12]	0.92	0.90	0.92	0.93	0.92	0.92	0.94	0.92	0.89	0.90
Document [13]	0.93	0.92	0.93	0.93	0.93	0.92	0.93	0.94	0.91	0.90
This paper	0.94	0.93	0.93	0.94	0.92	0.93	0.95	0.96	0.93	0.92

Select five of the images for output comparison. The comparison figure is shown in Fig. 6. The image in this paper has a good visual effect and a clear edge. Through the details, it can be seen that the denoising algorithm in this paper has a good effect and the details are processed in place, showing the image after denoising more clearly.

4 Epilogue

The algorithm of this paper adopts the structure of convolutional autoencoder, using coding layer and decoding layer structure to clearly divide the network into two parts. Among them, the coding layer uses multiple Inception modules for feature extraction, and the decoding layer improves the traditional Inception module, modifying the convolution network into a deconvolution network, so that the image can make full use of the advantages of Inception module feature extraction in the deconvolution network. Better integrate image features, restore original image information. From the experimental results, it can show good robustness in image denoising. However, there are also deficiencies. Compared with the convolutional neural network and the convolutional autoencoder without the Inception module, the four-layer network proposed by the algorithm of this paper takes a long time. After preliminary experimental tests, it is known that the use of the Inception module causes the network to become wider, and the volume The cumulative neural network has experienced more operations, so how to shorten the model training time is the focus of future research.

House1 Original Drawing   Noise Figure   Document[12]

Document [13]   Document [14]   This Paper

Woman2 Original Drawing   Noise Figure   Document [12]

Document [13]   Document [14]   This Paper

Camera Original Drawing   Noise Figure   Document [12]

Fig. 6. Rendering of each algorithm

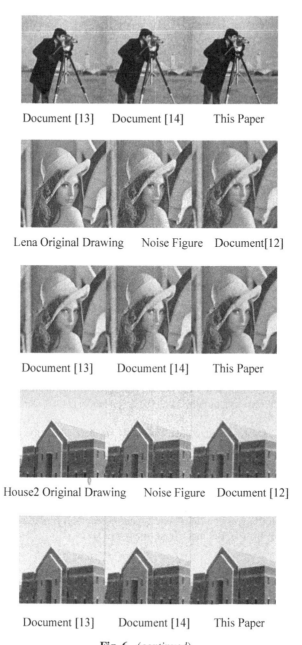

Fig. 6. (*continued*)

References

1. Gai, S., Bao, Z.Y.: New image denoising algorithm via improved deep convolutional neural network with perceptive loss. Exp. Syst. Appl. **138**, 112815 (2019)

2. Schuler, C.J., Christopher Burger, H., Harmeling, S.: A machine learning approach for non-blind image deconvolution. In: IEEE Conference on Computer Vision and Pattern Recognition, pp. 1067–1074 (2013)

3. Li, Y.J., Zhang, J., Wang, J.: Improved BM3D denoising method. IET Image Process. 11(12), 1197–1204 (2017)

4. Burger, H.C., Schuler, C.J., Harmeling, S. Image denoising: can plain neural networks compete with BA-I3D? In: Proceedings of the 2012 IEEE Conference on Computer Vision and Pattern Recognition (CVPR), Providence, RI, USA, pp. 2392–2399. IEEE (2012)

5. Chen, Y., Pock, T.: Trainable nonlinear reaction diffusion: a flexible framework for fast and effective image restoration. IEEE Trans. Pattern Anal. Mach. Intell. 39(6), 1256–1272 (2016)

6. Niu, W.H., Meng, J.L., Wang, Z.: Image denoising based on adaptive contraction function contourlet transform. J. Graph. 4, 17 (2015)

7. Hinton, G.E., Osinder, S., The, Y.W.: A fast learning algorithm for deep belief nets. Neural Comput. 18(7), 1527–1554 (2006)

8. Bengio, Y., Lamblin, P., Popovici, D., et al.: Greedy layerwise training of deep networks. In: Proceedings of the 20th Annual Conference on Neural Information Processing System, pp. 153–160 (2006)

9. Ma, H.Q., Ma, S.P., Xu, Y.L., et al.: Image denoising based on improved stacked sparse denoising autoencoder. Comput. Eng. and Appl. 54(4), 199–204 (2018)

10. Chen, Q., Pan, W.M.: Design and implementation of image denoising based on autoencoder. J. Xinjiang Normal Univ. (Nat. Sci. Ed.) 37(02), 80–85 (2018)

11. Li, C.P., Qin, P.L., Zhang, J.L.: Research on image denoising based on deep convolution neural network. Comput. Eng. 43(03), 253–260 (2017)

12. Zhang, K., Zuo, W., Chen, Y., et al.: Beyond a gaussian denoiser: residual learning of deep cnn for image denoising. IEEE Trans. Image Process. 26(7), 3142–3155 (2017)

13. Li, M., Zhang, G.H., Zeng, J.W., Yang, X.F., Hu, X.M.: Image denoising method based on convolution neural network combined with Inception model [J/OL]. Comput. Eng. Appl., 1–8 (2019)

14. Zhou, M.J., Liao, Q.: Knowledge push based OR attribute similarity. Comput. Eng. Appl. 47(32), 135–137 (2011)

15. Huang, Z., Li, Q., Fang, H.: Iterative weighted sparse representation for X-ray cardiovascular angiogram image denoising over learned dictionary. IET Image Process. 12(2), 254–261 (2018)

16. Xiang, Q., Pang, X.: Improved denoising auto-encoders for image denoising. In: 2018 11th International Congress on Image and Signal Processing, BioMedical Engineering and Informatics (CISP-BMEI), Beijing, China, pp. 1–9 (2018)

Detection and Segmentation of Graphical Elements on GUIs for Mobile Apps Based on Deep Learning

Rui Hu, Mingang Chen[✉], Lizhi Cai, and Wenjie Chen

Shanghai Key Laboratory of Computer Software Testing and Evaluating, Shanghai Development Center of Computer Software Technology, Shanghai, China
cmg@sscenter.sh.cn

Abstract. Recently, mobile devices are more popular than computers. However, mobile apps are not as thoroughly tested as desktop ones, especially for graphical user interface (GUI). In this paper, we study the detection and segmentation of graphical elements on GUIs for mobile apps based on deep learning. It is the preliminary work of GUI testing for mobile apps based on artificial intelligence. We create a dataset, which consists of 2,100 GUI screenshots (or pages) labeled with 42,156 graphic elements in 8 classes. Based on our dataset, we adopt Mask R-CNN to train the detection and segmentation of graphic elements on GUI screenshots. The experimental results show that the mAP value achieves 98%.

Keywords: Mobile apps · GUI dataset · Instance segmentation · Mask R-CNN

1 Introduction

Nowadays, mobile apps are becoming indispensable in our daily life and work. As there are a large number of apps available for users in online stores. It is significant for a competitive app to run smoothly as users expect. Mobile apps that often crash or bug-prone are likely to be quickly abandoned [1] and negatively evaluated [2] by users. Automation testing is one of the most efficient and reliable solutions to ensure the quality, avoiding high cost and low accuracy of manual testing. The GUI is the medium for most mobile apps to interact with users. This makes GUI testing an essential and prominent part of automation testing [3]. Obviously, the importance of GUI testing for mobile apps is self-evident. In this paper, the detection and segmentation technology of graphical elements on GUIs for mobile apps is developed on the basis of deep learning, which is the preliminary work of artificial intelligence-based GUI testing for mobile apps.

The traditional methods of GUI testing usually require testers to manually write a large number of test scripts. These tasks involve a lot of mechanical and repetitive work, and are time-consuming and laborious. Meanwhile, even small changes in GUI can destroy the entire test suite [4] and make the original test script invalid. Due to the

© ICST Institute for Computer Sciences, Social Informatics and Telecommunications Engineering 2020
Published by Springer Nature Switzerland AG 2020. All Rights Reserved
J. Liu et al. (Eds.): MobiCASE 2020, LNICST 341, pp. 187–197, 2020.
https://doi.org/10.1007/978-3-030-64214-3_13

large number of interactions on current GUIs and the increasing complexity of mobile apps, it is practically impossible to automate test generation with sufficient coverage. Muccini et al. [5] emphasized the main challenges of GUI testing for mobile apps: the large number of contextual events and responses, the diversity of devices and screen resolution, and the rapid updating of operating systems.

Today, with the improvement of computing performance, deep learning has made great progress in image detection. Based on Region convolution neural network (R-CNN) [6], the detection technology of graphical elements on GUIs is an end-to-end method. In which, the inputs of the network are the graphic elements on GUI images and their corresponding labels, and the outputs are the results of classification and localization.

In this paper, we apply the deep learning-based detection and segmentation technology to GUI pages. In traditional GUI testing, computers usually cannot recognize graphic elements. We humans can interact with GUIs according to their characteristics. We know that clicking the "Pay" button means paying, dragging the slider means controlling video progress, and sliding the screen means switching app pages. In our approach, the computer can intelligently identify the classification and localization of graphic elements on GUI pages just like human beings and operate the graphic elements without test scripts. To our knowledge, no one has yet applied the deep learning-based detection and segmentation algorithm to the GUI, which means that our work has great innovation.

The main contributions of this paper are summarized as follows:

(1) We create a dataset for detection and segmentation of graphic elements on GUI pages. 2,100 GUI screenshots in our dataset come from the Rico dataset [7] (a large-scale repository of GUI screenshots for mobile apps), Google Play and HUAWEI AppGallery. 42,156 graphic elements are manually labeled from GUI screenshots and classified into 8 classes depending on their types, appearances and operations imposed.

(2) The Mask R-CNN detection and segmentation algorithm is adopted to classify, locate and segment graphic elements on GUIs. Mask R-CNN is a multitask algorithm, which can be applied to train for classification, localization and masking together instead of training in stages. The experimental results show that the mean average precision (mAP) of detection and segmentation for graphic elements on GUI pages achieves 98%, which fully complies with the requirements of artificial intelligence-based GUI testing.

Section 2 gives a survey on the related work containing traditional GUI testing and object detection and segmentation algorithms. Section 3 presents two procedures of our work: the creation and manual labeling of our dataset, and the training and inferring of Mask R-CNN based on ResNet-50 network. Section 4 discusses experimental results and evaluation. Section 5 draws conclusions and prospects the future works.

2   Related Work

The significance of GUI testing for mobile apps is becoming self-evident due to the increasing popularity of mobile devices. Today, there are many approaches to test GUI for mobile apps. In [8], these techniques can be roughly divided into three categories: random-based testing [9, 10], model-based testing [11–13] and systematic testing [14, 15]. The random-based testing inputs random activity sequences to discover potential defects and bugs. For example, Monkey [9] sends random activities to random locations on the screen without considering the GUI structure. However, it is not suitable for large apps due to the large number of random behavior sequences that need to be processed in memory. Dynodroid [10], which is smarter than Monkey, maintains the input events through a context-sensitive approach. The model-based testing technology [16] builds detailed models of the relationship between events and GUI screens, and then generates test cases automatically according to these models. Based on the established model, MobiGUITAR [11] performs possible events by changing the state of the GUI. The systematic testing technique attempts to input specific test inputs according to pre-determined targets, such as exploring width or depth. A^{3}E [14] approach can explore apps systematically with Targeted Exploration technique and Depth-first Exploration technique.

Currently, mainstream object detection algorithms include Fast R-CNN [17], Faster R-CNN [18], YOLO [19], SSD [20] and YOLO9000 [21]. Object detection has enormous significance in many fields such as face detection [22] and video surveillance [23]. For a given image, Faster R-CNN returns the class label and bounding box coordinate of each object in the image. Mask R-CNN [24] is an extended instance segmentation algorithm of Faster R-CNN. Therefore, for the given image, Mask R-CNN will return segmentation masks of the object in addition to the class labels and bounding box coordinates. For Faster R-CNN, it uses VGG-16 [25] network to extract the feature map (FM) from the image. The region proposal network (RPN) [18] transmits these FMs and return candidate bounding boxes. Then the merging layer of region of interest (ROI) will be applied to these candidate bounding boxes to make all candidates have the same size.

Similar to the VGG-16 network in Faster R-CNN, from the images, Mask R-CNN uses the ResNet-101 [26] architecture to extract feature maps (FMs), which are regarded as inputs to next layer. Then, these FMs will be implemented to the region proposal network (RPN) to predict whether there are objects in the region. In this step, regions or FMs containing some objects can be predicted and obtained. The regions obtained from RPN may have different shapes, so the pooling layer will be applied to convert all regions to the same shape. These regions will be input to a fully connected layer to predict class labels and bounding boxes. Next, Mask R-CNN will generate segmentation masks [17]. For all predicted regions, the intersection over union (IOU) with the bounding boxes of real data can be calculated as follows:

$$IOU = \frac{\text{bounding box of candidate regin} \cap \text{bounding box of real data}}{\text{bounding box of candidate regin} \cup \text{bounding box of real data}}. \tag{1}$$

With the ROI based on the IOU value, the mask branch can be added to the existing architecture, which will return the segmentation mask for each region that contains the object.

3   Methodology

3.1   Dataset Creation

2,100 GUI screenshots in our dataset are selected from the Rico dataset, Google Play and HUAWEI AppGallery. The Rico [7] is a large crowd-sourced GUI interaction dataset which contains GUI screenshots and human interactions. Google Play and HUAWEI AppGallery are two online app stores for Android devices. Among our dataset, 1,600 GUI screenshots are provided from Rico dataset, 300 are provided from Google Play and 200 are provided from HUAWEI AppGallery.

Because the label and interaction data in Rico are not designed for our purpose, we first need to reclassify the graphic elements. Rico divides UI components into 25 classes. However, some components, such as Advertisement class, are sub-standard for our requirements. According to the types, appearances and actions applied of graphic elements, we integrate and divide components in Rico into 8 classes: Checkbox, Icon, Input, On/Off Switch, Page Indicator, Radio Button, Slider and Text Button.

Next, we manually label graphic elements on these 2,100 GUI screenshots. We label a total of 42,156 graphic elements and the label information includes their pixel-level coordinates, bounding boxes and classes. Figure 1 shows these 8 classes of graphic elements. Table 1 shows the quantity distribution of 8 graphic element classes. The number of icon and text button is huge since to their complexity and wide distribution.

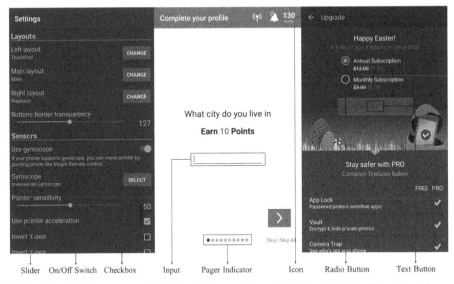

Slider   On/Off Switch   Checkbox   Input   Pager Indicator   Icon   Radio Button   Text Button

Fig. 1. Graphic elements are divided into 8 classes: Slider, On/Off switch, Checkbox, Input, Pager indicator, Icon, Radio button and Text button.

Table 1. Quantity distribution of 8 graphic element classes in our dataset.

Classification	Number
Checkbox	1,146
Icon	24,804
Input	1,912
On/off switch	1,566
Page indicator	1,032
Radio button	1,726
Slider	1,740
Text button	8,230

3.2 Mask R-CNN for Detection and Segmentation

Previously, the recognition of GUI graphic elements is based on source code. With Mask R-CNN, graphic elements on GUI pages can be identified intelligently by computers without source code.

Input. Our dataset is randomly divided into training (1,800 pages), validation (200 pages) and testing (100 pages) sets. Before training, the training and validation set should be preprocessed. GUI screenshots are converted to grayscale images to eliminate the influence of color on the convolutional neural network and speed up training. Then, the images have one channel and their size is 1440×2560.

Network Configuration. The ResNet-101 structure in Mask R-CNN is too deep to be quickly trained and deployed. Therefore, for our dataset, ResNet-101 is replaced by an agile ResNet-50 network. After considering the number of filtered ROI and the bounding box correlation between candidate region and real data, we choose the threshold value of IOU as 0.7. Due to the limitation of GPU memory, we set the batch size as 4. The momentum constant is set to 0.9 to accelerate the training. The learning rate is set to 0.0005 to prevent over fitting.

Training. Our Mask R-CNN network architecture is shown in Fig. 2. The labeled images are entered into ResNet-50 network to obtain the corresponding FMs. Each pixel in FMs has an anchor to obtain multiple candidate ROIs. These candidate ROIs are sent to RPN for binary classification (graphic elements or background) and candidate bounding box regression. Then the candidate ROIs will be filtered by IOU value. The predicted region is considered as ROI only when IOU value is greater than or equal to 0.7, otherwise it will be ignored. The process above is performed on all candidate regions. The remaining ROIs are operated by ROIAlign. Finally, the multitask loss L [17, 24] is applied on these ROIs for classification (8 classes of graphic elements and the background), bounding box regression and segmentation mask generation:

$$L = L_{cls} + L_{box} + L_{mask}. \tag{2}$$

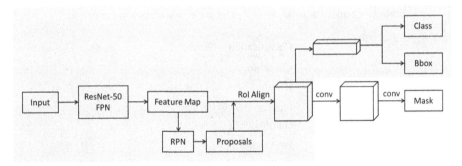

Fig. 2. Mask R-CNN architecture based on ResNet-50, in which the shadow part is Faster R-CNN architecture.

The classification loss L_{cls} is cross-entropy loss for object. The bounding box regression loss L_{box} is

$$L_{box}\left(t_i, t_i^*\right) = \sum_{i \in \{x,y,w,h\}} smooth_{L_1}\left(t_i - t_i^*\right), \tag{3}$$

in which

$$smooth_{L_1}(x) = \begin{cases} 0.5x^2 & if\ |x| < 1 \\ |x| - 0.5\ otherwise \end{cases}, \tag{4}$$

where t_i is predicted bounding box regression offsets and t_i^* is true offsets, $\{x, y, w, h\}$ is the center coordinate, width and height of the box. Based on the using of a per-pixel sigmoid, the mask loss L_{mask} is defined as average binary cross-entropy loss.

4   Experimental Results

4.1   Training Environment

The machine we used to train and test the Mask R-CNN model is a server with an NVidia Tesla P100 PCI-E GPU. The operating system of the machine is Ubuntu 16.04. The model is implemented with TensorFlow.

Since training all the networks will spend too much time and our dataset is small, we used a pre-trained model that was formed on the COCO dataset based on transfer learning. A total of 100 epochs were trained in the experiment, which took three days.

4.2   Evaluation

The used evaluation metric is the mean average precision (mAP), which is the average precision (AP) of all classes. The formula of mAP is as follows:

$$mAP = \frac{1}{|Q_R|} \sum_{q \in Q_R} AP(q). \tag{5}$$

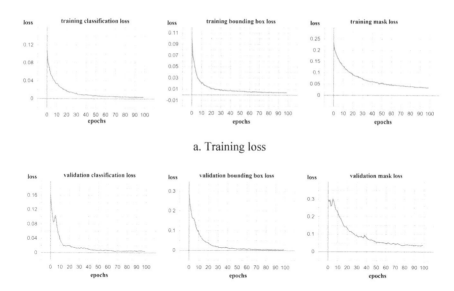

a. Training loss

b. Validation loss

Fig. 3. The consequences of loss in training and validation processes, including classification loss, bounding box regression loss and mask loss.

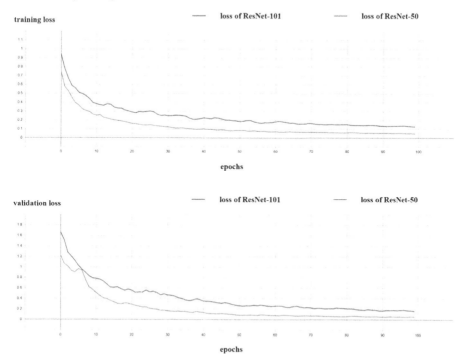

Fig. 4. The training and validation overall loss comparison of Mask R-CNN trained by ResNet-50 and ResNet-101.

Fig. 5. Detection and segmentation results for some images. The label classes, bounding boxes and segmentation masks for graphic elements are shown.

Where AP(q) is the average precision of each class and Q_R is the number of classifications. AP value is the area under the curve of precision and recall. For testing, the experimental results demonstrate that the mAP value of ResNet-50 is 98%. The training and validation loss are shown in Fig. 3. In addition, on the basis of using our dataset, we compare two backbones networks of Mask R-CNN, ResNet-50 and ResNet-101. The overall loss is the sum of classification loss, bounding box loss and mask loss. ResNet-50 achieves 0.047 overall loss on training dataset and 0.049 on validation dataset. While the overall loss of ResNet-101 on training dataset is 0.13 and that on validation dataset is 0.15. Figure 4 shows the comparison of overall loss trained by ResNet-50 and ResNet-101.

Figure 5 shows the results of some images that contain the label classes, bounding boxes and segmentation masks of graphic elements. The results show that the Mask R-CNN based on ResNet-50 can brilliantly recognize graphic elements on GUIs.

5 Conclusion

In this paper, we create a dataset consists of 2,100 GUI screenshots for detection and segmentation tasks of graphical elements on GUIs. 1,600 screenshots are selected from Rico dataset, 300 are selected from Google Play and 200 are selected from HUAWEI AppGallery. The GUI screenshots are manually labeled a total of 42,156 graphic elements, which are divided into 8 classes according to their types, appearances and actions applied. In addition, Mask R-CNN based on ResNet-50 is applied to the detection and segmentation of graphic elements on GUIs. The mAP value achieves 98%, showing the brilliant application effect. As for future work, on the basis of this paper, we will study GUI testing based on artificial intelligence. We will also improve our method to support more complex events such as context events and capturing GUI bugs.

Acknowledgement. This work is funded by National Key R&D Program of China (No. 2018YFB1403400), and Science and Technology Commission of Shanghai Municipality Program, China. (Nos. 17411952800, 18DZ2203700, 18DZ1113400).

References

1. Moran, K., Linares-Vasquez, M., Bernal-Cardenas, C.: Automatically discovering, reporting and reproducing android application crashes. In: ICST, pp. 33–44. IEEE Computer Society, Los Alamitos (2016)
2. Khalid, H., Shihab, E., Nagappan, M.: What do mobile app users complain about? IEEE Softw. **32**(3), 70–77 (2015)
3. Kaur, A.: Review of mobile applications testing with automated techniques. Int. J. Adv. Res. Comput. Commun. Eng. **4**(10), 503–507 (2015)
4. Coppola, R., Raffero, E., Torchiano, M.: Automated mobile UI test fragility: an exploratory assessment study on Android. In: INTUITEST 2016: Proceedings of the 2nd International Workshop on User Interface Test Automation, pp. 11–20. ACM, New York (2016)
5. Muccini, H., Francesco, A.D., Esposito, P.: Software testing of mobile applications: challenges and future research directions. In: International Workshop on Automation of Software Test (AST), pp. 29–35. IEEE Computer Society, Los Alamitos (2012)

6. Girshick, R., Donahue, J., Darrell, T., Malik, J.: Rich feature hierarchies for accurate object detection and semantic segmentation. In: CVPR, pp. 580–587. IEEE Computer Society, Los Alamitos (2014)

7. Deka, B., Huang, Z., Franzen, C.: Rico: a mobile app dataset for building data-driven design applications. In: UIST '17: Proceedings of the 30th Annual ACM Symposium on User Interface Software and Technology, pp. 845–854. ACM, New York (2017)

8. Choudhary, S., Gorla, A., Orso, A.: Automated test input generation for android: are we there yet? In: 30th IEEE/ACM International Conference on Automated Software Engineering, pp. 429–440. IEEE Computer Society, Los Alamitos (2015)

9. UI/Application Exerciser Monkey. https://developer.android.com/studio/test/monkey.html Accessed 27 July 2020

10. Machiry, A., Tahiliani, R., Naik, M.: Dynodroid: an input generation system for android apps. In: Proceedings of the 2013 9th Joint Meeting on Foundations of Software Engineering, pp. 224–234. ACM, New York (2013)

11. Amalfitano, D., Fasolino, A., Tramontana, P., Ta, B., Memon, A.: MobiGUITAR: automated model-based testing of mobile apps. IEEE Softw. **32**(5), 53–59 (2015)

12. Yang, W., Prasad, M.R., Xie, T.: A grey-box approach for automated GUI-model generation of mobile applications. In: Cortellessa, V., Varró, D. (eds.) FASE 2013. LNCS, vol. 7793, pp. 250–265. Springer, Heidelberg (2013). https://doi.org/10.1007/978-3-642-37057-1_19

13. Amalfitano, D., Fasolino, A., Tramontana, P., De Carmine, S., Memon, A.: Using GUI ripping for automated testing of android applications. In: Proceedings of the 27th IEEE/ACM International Conference on Automated Software Engineering, pp. 258–261. ACM, New York (2012)

14. Azim, T., Neamtiu, I.: Targeted and depth-first exploration for systematic testing of android apps. In: Proceedings of the 2013 ACM SIGPLAN International Conference on Object Oriented Programming Systems Languages & Applications, pp. 641–660. ACM, New York (2013)

15. Bhoraskar, R., Han, S., Jeon, J.: Brahmastra: driving apps to test the security of third-party components. In: Proceedings of the 23rd USENIX Conference on Security Symposium, pp. 1021–1036. USENIX Association, San Diego (2014)

16. Pretschner, A., Prenninger, W., Wagner, S.: One evaluation of model-based testing and its automation. In: Proceedings of the 27th International Conference on Software Engineering, pp. 392–401. ACM, New York (2015)

17. Girshick, R.: Fast R-CNN. In: ICCV, pp. 1440–1448. IEEE Computer Society, Los Alamitos (2015)

18. Ren, S.Q., He, K.M., Girshick, R.B.: Faster R-CNN: towards real-time object detection with region proposal networks. In: NIPS, pp. 91–99. MIT Press, Cambridge (2015)

19. Redmon, J., Divvala, S., Girshick, R.: You only look once: unified, real-time object detection. In: CVPR, pp. 779–788. IEEE Computer Society, Los Alamitos (2016)

20. Liu, W., et al.: SSD: single shot multibox detector. In: Xie, T., Leibe, B., Matas, J., Sebe, N., Welling, M. (eds.) ECCV 2016. LNCS, vol. 9905, pp. 21–37. Springer, Cham (2016). https://doi.org/10.1007/978-3-319-46448-0_2

21. Redmon, J., Farhadi, A.: YOLO9000: better, faster, stronger. In: CVPR, pp. 6517–6525. IEEE Computer Society, Los Alamitos (2017)

22. Wei, Z.Y., Wen, C., Xie, K.: Real-time face detection for mobile devices with optical flow estimation. J. Comput. Appl. **38**(4), 1146–1150 (2018)

23. Li, J.W., Zhou, X.L., Chan, S.X.: A novel video target tracking method based on adaptive convolutional neural network feature. J. Comput. Aided Des. Comput. Graph. **30**(2), 273–281 (2018)

24. He, K.M., Gkioxari, G., Dollar, P.: Mask R-CNN. In: ICCV, pp. 2980–2988. IEEE Computer Society, Los Alamitos (2017)

25. Simonyan, K., Zisserman, A.: Very deep convolutional networks for large-scale image recognition. arXiv.org, https://arxiv.org/abs/1409.1556 Accessed 3 July 2020
26. He, K.M., Zhang, X.Y., Ren, S.Q.: Deep residual learning for image recognition. In: CVPR, pp. 770–778. IEEE Computer Society, Los Alamitos (2016)

Approximate Sub-graph Matching over Knowledge Graph

Jiyuan Ren$^{(\boxtimes)}$, Yangfu Liu, Yi Shen, Zhe Wang, and Zhen Luo

Northeast Branch of State Grid Corporation of China, Liaoning, China
`582854926@qq.com`

Abstract. With the rapid development of the mobile internet, the volume of data has grown exponentially, and the content of data become more complicated. It is hard for people to select useful information from such a large number of data. In this paper, we study the problem of approximate sub-graph matching over knowledge graph. We first propose two algorithms to reduce the scale of knowledge graph. Next, we use an efficient algorithm to find similarity sub-graphs. Thirdly, we use skyline technique to further select high quality sub-graphs from the matching results. Theoretical analysis and extensive experimental results demonstrate the effectiveness of the proposed algorithms.

Keywords: Knowledge graph · Sub-graph matching · Compression · Skyline

1 Introduction

With the rapid development of the mobile internet [1,2], the volume of data has grown exponentially, and the content of data become more complicated. It is hard for people to select useful information from such a large number of data. In addition, when the target of a user is an object- portfolio other than a signal object, the problem turns to more difficult since users should fully consider relationships among different data.

For example, in a event-based mobile social network [3,4], users often select object-portfolio, i.e., shop, sports, and cinemas based on the their location, their business hours and so on. For example, Peter usually attends sports activities organized on Meetup. In the evening of Friday, Peter faces a dilemma since Meetup recommends four three conflicting sport activities on Sunday: a hiking trip from 7:00 a.m. to 11:00 p.m. (at the place A), a badminton game from 10:00 a.m. to 14:00 a.m (at the place D), a basketball game from 11:30 a.m. to 2:30 p.m. (at the place F), and a football game from 1:30 a.m. to 3:30 p.m. (at the place G). Though Peter is interested in all these four sports, he can only attend few of them. Thus, Peter must make a decision to select an object- portfolio.

In a network attack defense system, the system should process data package generated from a large number of IP addresses, and it has to identify malicious network attack and attack mode. Usually, the attack mode is complex. As an example, a compound attack often involves multiple attacks step, we use could use a graph structure to express the relationship among attackers and hosts.

© ICST Institute for Computer Sciences, Social Informatics and Telecommunications Engineering 2020
Published by Springer Nature Switzerland AG 2020. All Rights Reserved
J. Liu et al. (Eds.): MobiCASE 2020, LNICST 341, pp. 198–208, 2020.
https://doi.org/10.1007/978-3-030-64214-3_14

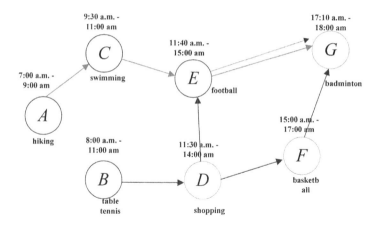

Fig. 1. Event-based mobile social network

However, compared with simple graph structure [5–7], examples discussed above is very complex. The main reason is we should associate each nodes with many information. Back to the example in Fig. 1. We should maintain the starting (also ending) moment for each node. In addition, we should maintain the keywords for each node. Thus, we introduce knowledge graph (short for **KG**). A KG is a heterogeneous graph [8–10], where nodes function as entities, and edges represent relations between entities. Items and their attributes can be mapped into the **KG** to understand the mutual relations between items. Moreover, users and user side information can also be integrated into **KG**. The benefit is users can accurately capture relationship among users and items.

In this paper, we study the problem of subgraph matching over large scale knowledge graph. Since it is a NP-hard problem, in order to solve this problem, we propose an approximate framework named **CBSM** (short for Compress-based Sub-graph Matching). Given a graph $G(V, E)$, we first compress it to a sub-graph G'. Compared with G', both the edge scale and vertex scale are all reduced a lot. In this way, we achieve the goal of using a small number of objects to support query processing. Next, we propose a skyline-based sub-graph matching algorithm. Compared with other algorithms, it considers both the similarity of graph structure and the similarity of key words. Above all, the contribution of this paper is as follows.

- We propose a novel algorithm to compress knowledge graph. It first computes the weight of all edges. Based on the computing result, we propose a randomly algorithm to delete edges (also vertexes) that are not important. Since the scale of the knowledge graph is reduced, the cost of the subgraph matching algorithm also could be reduced.
- We propose a skyline-based algorithm to find similarity sub-graphs from knowledge graph G. It evaluates the similarity via the following two facets. Firstly, it evaluates the similarity based on the graph structure. Secondly, it evaluates the similarity based on key word set difference between query graph and sub-graphs in G.

The rest of this paper is organized as follows: Sect. 2 gives background, Sect. 3 presents our proposed solution. Section 4 evaluates the proposed methods with extensive experiments. Section 5 is the conclusion.

2   Background

In this section, we first review the algorithms about sub-graph matching. Thereafter, we explain the problem definition (Table 1).

Table 1. The summary of notations

Notation	Definition
KG	A knowledge graph
V	The vertex set
E	The edge set
K	The key word set
Q_S	The skyline sub-graph
KG_C	The compressed knowledge graph

2.1   Related Work

Matching over Static Graph. Many researchers studied the problem of static graph matching. Lee et al. compared five classic static graph matching algorithms under the same environment, i.e., vF2, quicksi, graphql, Gaddi, sPath and so on. They explain the advantages and disadvantages of these five algorithms. Based on the application types, Yu Jing et al. classify graph matching based on different application, such as structural matching, semantic matching, precise matching, approximate matching as well as optimal algorithm and approximate algorithm. It focuses on the algorithms about static graph matching over static graph. However, it is not suitable to solve the problem of pattern matching in dynamic graph data.

Matching over Dynamic Graph. Many researchers studied the problem of sub-graph matching over dynamic graph. They mainly focus on the incremental maintenance algorithm when a new vertex (or a new edge) is inserted into the graph. Wang et al. propose a neighbour-tree based method to support subgraph matching over dynamic graph. Their key idea is using neighbour-tree to filter sub-graphs that have no chance to become query results. The benefit is the number of sub-graphs that should be verified could be effectively reduced.

With the emergence of data center, many researches use network topology structure to express relationships among different centers. Compared with traditional dynamic graph, the topological structure of graph data will not change frequently. However, the attribute values of nodes may be changed frequently. In order to process this problem, zhong et al. proposed the gradin algorithm. The key idea behind it is it uses the attribute value of each node in the frequent subgraph as a data dimension. In this way, the frequent subgraph can be represented as an n-dimensional coordinate vector in the n-dimensional grid index. In this way, all frequent subgraphs of the data graph can be indexed.

In many cases, the system has the requirement of real time. Thus, approximate algorithms with error guarantee is more preferred in many cases. Among all the algorithms, IncIsoMatch, SJ$®Tree and Gradin are representative ones. Their key idea is using relationship among vectors to approximately evaluate similarity between query graph and sub-graphs. Thus, the complex of these algorithms are reduced a lot.

2.2   Problem Definition

In this section, we first explain the concept of Dominance. Thereafter, we formally discuss the problem definition. For simplicity, let o be an object in $d-$dimensional space. Its coordinate is expressed the tuple $\langle o[1], o[2], \ldots, o[d] \rangle$. Without loss of generality, we assume that a smaller value indicates a better performance in all dimensions. Assuming $o_1[i]$ refers to the value of o_1 in the i-th dimension, we say $o_1 \prec o_2$ if $\forall i o_1[i] \leq o_2[i]$ and $\exists j o_1[j] \leq o_2[j]$.

Definition 1 DOMINANCE. *Let $\mathcal{D}$ be a set of $d-$dimensional objects with size N. $\forall o_1, o_2 \in \mathcal{D}$, we say object o_1 dominates o_2 (denoted as $o_1 \prec o_2$) if o_1 is as good as or better than o_2 in all dimensions, and better in at least one dimension.*

Definition 2 SKYLINE. *For each object $o \in \mathcal{D}$, if o is not dominated by any objects in $\mathcal{D}$, we regard o as a skyline object. All skyline objects in $\mathcal{D}$ make up of the skyline set, i.e., denoted as $\mathcal{S}$.*

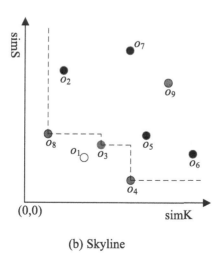

(b) Skyline

Fig. 2. An example of skyline

As is shown in Fig. 2, there are 9 2-dimensional objects. Among them, o_3, o_4 and o_8 are skyline sub-graphs, we regard them as skyline objects. After discussing the concept of dominate and skyline, we now explain the concept of RDF graph and sub-graph matching over RDF graph.

Definition 3 KNOWLEDGE GRAPH. *Let E, R and S be three sets. They refer to the URI, blank nodes and literal. We call a tuple $\langle s, p, o \rangle \in (U \cup B) \times (U \cup B \cup L)$. Here, s is the subject, p refers to the predicate, and o refers to the object.*

Definition 4 SUB-GRAPH MATCHING OVER KNOWLEDGE GRAPH. *Let KG be a RDF Graph, $q\langle K, G \rangle$ be a query graph. q search on KG, find a sub-graph G' satisfying that the key word set of G' equals to $q.K$. In addition, G' should equals to $q.G$.*

Definition 5 APPROXIMATE SUB-GRAPH MATCHING OVER KNOWLEDGE GRAPH. *Let KG be a RDF Graph, $q\langle K, G \rangle$ be a query graph. q search on KG, find a sub-graph G' satisfying the following two cases. Here, $\mathsf{simK}(q.K, G'.K)$ refers to the similarity between keyword set of q and G'. $\mathsf{simS}(q.S, G'.S)$ refers to the similarity between structure of q and G'.*

- $\mathsf{simK}(q.K, G'.K) \leq \varepsilon$
- $\mathsf{simS}(q.S, G'.S) \leq \tau$

We find that, in real applications, it is difficult to find such a perfect sub-graph. In addition, the computational cost is also high. In most cases, we can find the solution in real time. Thus, in this paper, we propose the approximate RDF Sub-graph Matching over RDF Graph. We now formally discuss the problem definition.

Definition 6 SKYLINE BASED SUB-GRAPH MATCHING. *Let KG be a RDF Graph, $q\langle K, G \rangle$ be a query graph. q search on KG, find a sub-graph KG' satisfying that $\mathsf{sim}(q.K, KG'.K)$ is smaller than ε. In addition, $\mathsf{sim}(q.G, KG'.G)$ is smaller than η. Let $\mathcal{G}$ be a set of sub-graphs satisfying the above conditions, the find result set $\mathcal{G}$ is the skyline sub-graphs among all the sub-graphs in $\mathcal{G}$.*

3 Skyline Based Sub-RDF Matching

In this section, we propose a novel framework to support skyline based Sub-RDF matching. First of this section, we explain how to compress a given RDF. Next, we discuss the approximate matching algorithm. Last of this section, we discuss the skyline algorithm.

3.1 The Compression Algorithm

Recalling Sect. 2.2, a given KG could be expressed by the tuple $\langle E, R, S \rangle$. Since both the set E and R contains multitudes of key words, the space cost of a KG is high. Lucky, we find that some key words are similar with each other, we could use one key word to express them. In addition, some entries are all similar with each other, we also could use one entry to express it.

Algorithm 1: The Compression Algorithm

Input: Node Set $\mathcal{E}$, Knowledge Graph KG
Output: Compressed Knowledge Graph KG'

1 **while** E *is not empty* **do**
2   Node $e \leftarrow E[0]$;
3   **for** i *from 2 to* $|E|$ **do**
4   **if** *sim(e, E[i])* $\geq \zeta$ **then**
5              $M \leftarrow E[i]$;
6              $E \leftarrow E - E[i]$;
7      $E' \leftarrow$ construction(M);

8 **for** i *from 1 to* $|E|$ **do**
9   Node $e \leftarrow E[i]$;
10   **for** j *from 1 to* $|e|$ **do**
11         $sum \leftarrow sum + e[j].w$;
12   **for** j *from 1 to* $|e|$ **do**
13         $I[j] \leftarrow \frac{e[j].w}{sum}$;
14   **if** $I[j] \leq \tau$ **then**
15   delete$(e, e[j])$;
16   **if** $|e[j]| = 0$ **then**
17   delete$(e[j])$;

18 return;

Take an example in Fig. 2. The node A, B, and C are three basketball venues. The service provided by them are expressed a group of key words. Since the key word set of them are similar with each other, we could construct a super node U. It contains the union of A, B, C's key words. Accordingly, we could use the nodes $U - A$, $U - B$, and $U - C$ to express nodes A, B, and C. Compared with the orient ones, the space cost could be reduced a lot (Fig. 3).

In order to compress key word information of each node, for each node $e \in E$, we compute the similarity of e and all the other nodes. After computing, we select all the ones whose similarity to e is smaller than a threshold ζ, construct a super node based on e and these nodes, and remove these nodes from E.

From then on, we repeat the above operations to process other nodes. The above operations is terminated until E turns to empty. Besides, we also should delete edges (vertexes) which importance is low. For example, given a vertex v and a group of other vertexes $\{v_1, v_2, \ldots, v_m\}$, for each i, it satisfied that there exists an edge $e(v, v_i)$. We evaluate the importance of an edge based on its weights. It can be computed based on Eq. 1.

$$\mathbf{I}_i = \frac{|w_i|}{\sum_{i=j}^{i=m} |W_j|} \tag{1}$$

It implies if the importance of an edge is high, the weight of this edge is high. Otherwise, the weight is low. Thus, if the importance of edge is smaller than a threshold τ, we could delete it directly. Based on the about observations, we could access every edge to compute its weight, and delete the ones whose importance is smaller than a threshold τ. In particularly, if a vertex has no edge, this vertex could be deleted directly.

As is shown in Algorithm 2, it contains two steps. In the first step, we access every node $e \in E$, find other nodes which are similar with e. After finding, we construct a super node based on these nodes. From then on, we repeat the above operations to process other ones. In the second step, we access every node $e \in E$, compute the weight sum of all vertexes that are connected with e. After computing, we delete all unimportant ones. In particularly, if we find another node e' which has no edge, we delete it. This algorithm is terminated until processing all nodes.

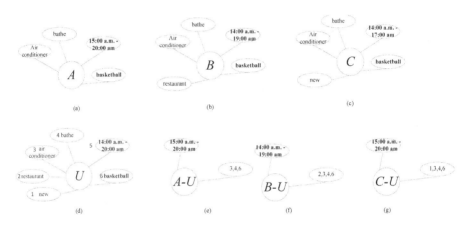

Fig. 3. Key word based compression

3.2 The Sub-graph Matching Algorithm

Given a query graph $G\langle S, K \rangle$, we first find the enumerate sequence. Since the enumeration algorithm have been discussed in many algorithms, we skip the details for saving space. Let CG be a set of enumerated result. These sub-graphs contain two types, that are, connected-graph and unconnected-graph. If a sub-graph is a unconnected-graph, it must not be a query result. Otherwise, we should further process it.

When a graph $CG[i]$ is a connected-graph, the matching contains two types. The first one is we could not find any prefix for q. The second one is we could find a few prefixes for q. Under case one, since q has at least one spanning tree, we access these spanning trees to find similar sub-graph, and insert all the query result into a set M. In other words, if we can find a sub-graph SG satisfying that $\mathsf{sim}(SG, q) \leq \delta$, we insert it into query result set. Under case two, if we can find the matching for the prefix of q, assuming the query graph q is the longest prefix, we check whether $e_1, e_2, \ldots e_m$ are

---

Algorithm 2: The Sub-Graph Matching Algorithm

Input: Knowledge Graph $\mathcal{KG}$, query graph q
Output: Similarity Knowledge Graph Set SG

1 $T \leftarrow$ gRandomSpanningTree(KG);
2 $T \leftarrow$ nextsorting(T);
3 $T \leftarrow$ jump-sorting(T);
4 $\mathcal{L}\ CS \leftarrow$ Index-Matching(T);
5 **for** j *from 1 to* $|CS|$ **do**
6 $\quad \theta \leftarrow$ simK($CS[i], q$);
7 $\quad$ **if** $\theta \geq \varepsilon$ **then**
8 $\quad \quad \lfloor\ CS \leftarrow CS - CS[i];$

9 $\mathcal{L}$ return;

---

contained in the graph G. If the answer is yes, it must have a matching. From then on, we repeat the above operations to find other sub-graphs.

After finding all these similarity sub-graphs, we then access their key word set. Our goal is to a group of sub-graphs whose corresponding key word set is also similar with that of the query graph. Here, we evaluate the relevance between the query q and the key word set of a graph via using Eq. 1. Here, $q(d)$ refers to a key words set of q. In order to achieve this goal, we scan all these sub-graphs, compute the key words similarity among these sub-graphs and the query graph. Last of all, we select sub-graphs satisfying that simK$(g, q) \leq \varepsilon$.

$$simK(q.K, G.K) = \frac{|V(q.K) \cap q(q.K)|}{|q(G.K)|} \tag{2}$$

3.3 The Skyline Algorithm

Let CG be a set of sub-graphs. For each element $g \in CG$, it should satisfies that simK$(q, g) \leq \varepsilon$ and simS$(q, g) \leq \tau$. However, if we can find multitudes of sub-graphs from G, users still cannot select a property sub-graph. The reason behind it is it is difficult to find a suitable parameters δ and ε. In this section, we propose a skyline-based algorithm to solve this problem. Since the skyline query processing algorithms have been well studied, we skip the details for saving space.

4 Experimental Study

4.1 Experiment Setup

In this section, we use a real data set named **Meetup** as real data set. In this dataset, each node is associated with a set of tags and a location. Each event in the dataset is also associated with a location. They are used as vertexes. In addition, we use the reachability among different vertexes to construct edges. For example, given two event v_1 and v_2, there starting moment/end moment are [9.00pm, 11.pm], [9.30pm, 12.pm] respectively. Since the ending time of v_1 is later than the starting time of v_2, we cannot construct an edge between v_1 and v_2. By contrast, let v_3 be another event. Its starting moment/ending moment is [12.00pm, 2.am]. Thus, we construct an edge between v_1 and v_3. We also use synthetic dataset for evaluation. We generate the utility values following Uniform, Normal and Power distributions respectively. The algorithms are implemented in C++, and the experiments were performed on a Windows 10 machine with Intel i7-7600 3.40 GHz 8-core CPU and 32 GB memory (Table 2).

Table 2. Summary of datasets

Dataset	Description	Vertex amount	Edge amount
MEETUP	Vancouver	225	2012
SYN-U	Synthetic data set	364	3367

For the parameter setting, we first evaluate the parameters τ and ε to the algorithm. In addition, we should evaluate the impaction of query graph scale to the algorithm performance. The parameter setting is shown in Table 3.

Table 3. Summary of parameters

Parameter	Description	Vertex amount
τ	Key word similarity	0.1, 0.15, 0.2, 0.25, 0.3
ε	Graph structure similarity	0.1, 0.15, 0.2, 0.25, 0.3
$|q|$	The number of vertex	2, 5, 10, 20, 30

4.2 Algorithm Performance

In this section, we compare the framework CBSM with a baseline algorithm. Compared with CBSM, it does not compress the graph. In addition, it does not terminate the searching the number of query results achieves to 100. In this paper, we call it as OSM. In the following, we first compare their performance under different parameter τ.

As is depicted in Fig. 4, CBSM performs best of all. The reason behind is, for one thing, we use two compression algorithms to reduce the space cost of knowledge graph. Thus, we only need to spend lower space cost in I/O. For another, since the graph scale turns to much smaller than the orient graph, the computational cost also could turn to small. For the stability, we compare CBSM with CBSM under synthetic dataset.

As is depicted in Fig. 5, CBSM still performs best of all. Similar with the reason discussed before, we use two compression algorithms to reduce the space cost of knowledge graph. Another reason is, with the increasing of ε, it become easy to find subgraphs that satisfy the query conditions. Since CBSM stop searching when the number of query results achieve to 100, the running time of CBSM become lower when ε is high enough.

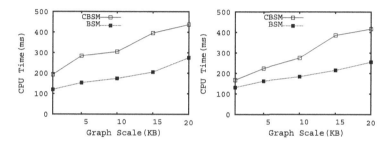

Fig. 4. Running time comparison under different graph scale

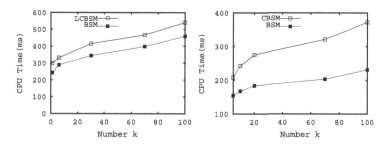

Fig. 5. Running time comparison under different ε

5   Conclusions

In this paper, we propose a novel framework named **CBSM** to support approximate sub-graph matching over knowledge graph. We first propose a compression algorithm to reduce the cost spent by keyword set. Next, we propose a weight based algorithm to delete unimportant edges (also vertexes). Based on these two algorithms, the scale of knowledge graph could be effectively reduced. Next, we use an efficient algorithm to find similarity sub-graphs and a skyline based method to select high quality sub-graphs. Theoretical analysis and extensive experimental results demonstrate the effectiveness of the proposed algorithms.

References

1. Amer-Yahia, S., Roy, S.B., Chawlat, A., et al.: Group recommendation: semantics and efficiency. Proceedings of the Vldb Endowment **2**(1), 754–765 (2010)
2. Bar-Yehuda, R., Bendel, K., Freund, A., Rawitz, D.: Local ratio: a unified framework for approximation algorithms. ACM Comput. Surv. (CSUR) **36**(4), 422–463 (2004). 1935–2004
3. Chen, C., Zhang, D., Guo, B., Ma, X., Pan, G., Wu, Z.: TripPlanner: personalized trip planning leveraging heterogeneous crowdsourced digital footprints. IEEE Trans. Intell. Transp. Syst. (T-ITS) **16**, 1259–1273 (2014)
4. Du, R., Yu, Z., Mei, T., Wang, Z., Wang, Z., Guo, B.: Predicting activity attendance in event-based social networks: content, context and social influence. In: UbiComp 2014: Proceedings of the 2014 ACM International Joint Conference on Pervasive and Ubiquitous Computing, pp. 425–434 (2014)

5. Feng, K., Cong, G., Bhowmick, S.S., Ma, S.: In search of influential event organizers in online social networks. In: SIGMOD 2014: Proceedings of the 2014 ACM SIGMOD International Conference on Management of Data, pp. 63–74 (2014). https://doi.org/10.1145/2588555.2612173

6. Wang, H., et al.: RippleNet: propagating user preferences on the knowledge graph for recommender systems. In: SIGMOD 2018, pp. 417–426

7. Bollacker, K., Evans, C., Paritosh, P., Sturge, T., Taylor, J.: Freebase: a collaboratively created graph database for structuring human knowledge. In: SIGMOD 2008, pp. 1247–1250

8. Suchanek, F.M., Kasneci, G., Weikum, G.: Yago: a core of semantic knowledge. In: World Wide Web 2007, pp. 697–706

9. Huang, X., Zhang, J., Li, D., Li, P.: Knowledge graph embedding based question answering. In: Proceedings of the Twelfth ACM International Conference on Web Search and Data Mining, pp. 105–113. ACM (2019)

10. Belleau, F., Nolin, M.-A., Tourigny, N., Rigault, P., Morissette, J.: Bio2RDF: towards a mashup to build bioinformatics knowledge systems. knowledge systems. J. Biomed. Inform. **41**(5), 706–716 (2008)

Metamorphic Testing for Plant Identification Mobile Applications Based on Test Contexts

Hongjing Guo[1], Chuanqi Tao[1,2,3(✉)], and Zhiqiu Huang[1,2]

[1] College of Computer Science and Technology, Nanjing University of Aeronautics and Astronautics, Nanjing 211100, China
{guohongjing,taochuanqi,zqhuang}@nuaa.edu.cn
[2] Ministry Key Laboratory for Safety-Critical Software Development and Verification, Nanjing University of Aeronautics and Astronautics, Nanjing 211100, China
[3] National Key Laboratory for Novel Software Technology, Nanjing University, Nanjing 210023, China

Abstract. With the fast growth of artificial intelligence and big data technologies, AI-based mobile apps are widely used in people's daily life. However, the quality problem of apps is becoming more and more prominent. Many AI-based mobile apps often demonstrate inconsistent behaviors for the same input data when context conditions are changed. Nevertheless, existing work seldom focuses on performing testing and quality validation for AI-based mobile apps under different context conditions. To automatically test AI-based plant identification mobile apps, this paper introduces TestPlantID, a novel metamorphic testing approach based on test contexts. First, TestPlantID constructs seven test contexts for mimicking contextual factors of plant identification usage scenarios. Next, TestPlantID defines test-context-based metamorphic relations for performing metamorphic testing to detect inconsistent behaviors. Then, TestPlantID generates follow-up images with various test contexts for testing by applying image transformations and photographing real-world plants. Moreover, a case study on three plant identification mobile apps shows that TestPlantID could reveal more than five thousand inconsistent behaviors, and differentiate the capability of detecting inconsistent behaviors with different test contexts.

Keywords: Test context · AI-based software · Metamorphic testing

1 Introduction

With the development of the mobile Internet, cloud computing, big data technologies as well as intelligent algorithms, there is a significant increase in

Supported by National Key R&D Program of China (2018YFB1003900), National Natural Science Foundation of China (61602267, 61402229), Open Fund of the State Key Laboratory for Novel Software Technology (KFKT2018B19), Fundamental Research Funds for the Central Universities (NO. NS2019058), and China Postdoctoral Science Foundation Funded Project (No. 2019M651825).

J. Liu et al. (Eds.): MobiCASE 2020, LNICST 341, pp. 209–223, 2020.
https://doi.org/10.1007/978-3-030-64214-3_15

mobile apps. Mobile apps are widely used in people's daily life including business, education, biomedical industry, social media, transportation, etc. Many software products based on artificial intelligence (AI) techniques emerge in the mobile app store, such as object recognition apps, navigation apps, and translation apps. AI-based apps are developed based on advanced Machine Learning (ML) algorithms through large-scale training data, which undoubtedly brings many new difficulties to the quality assurance and verification [1], such as defect analysis, defect prediction, and testing. It also puts forward a large market demand and research demand for quality assurance and verification.

Different from traditional software, AI-based software learns decision its logic from large-scale training data [2]. They are characterized by uncertainty and probabilities, dependence on big data, and constant self-learning from past behaviors [3]. Moreover, AI-based apps usually involve context issues, such as scenario, location [4], time, and stakeholders. Many AI functions of apps generate inconsistent outputs for the same input data when context conditions are changed [5]. For example, Fig. 1 shows two images of the same plant taken by a tester from different angles. For Fig. 1a, it was taken at an approximately 90-degree angle from the plane of the smartphone to the plant. For Fig. 1b, it was taken at an approximately 45-degree angle from the plane of the smartphone to the plant. A plant identification app named PlantSnap, identified Fig. 1a as a Lactuca virosa. However, it identified Fig. 1b as a Hydrangea macrophylla. Interestingly, different photographing angles of the same plant can even result in quite inconsistent behaviors. Therefore, the evaluation of the reliability and robustness of AI-based apps under different context conditions becomes an important task.

(a)   (b)

Fig. 1. Two images with different photographing angles of a plant

Besides, testing AI-based software has the oracle problem [6]. Currently, metamorphic testing (MT) has been successfully used in alleviating the test oracle problem [7]. Central to MT is a set of metamorphic relations (MRs), which depict the relationships between the results of multiple inputs and their expected output [7]. Specifically, for a plant identification app, the MRs are defined such that no matter how the context conditions are changed, the identification results are supposed to be consistent with the plant images under the original context conditions. Nevertheless, as a typical AI-based mobile app, there is a lack of approaches focusing on leveraging MRs based on different context conditions to test AI-based plant identification apps.

To address those issues, we propose a novel testing approach, namely Test-PlantID, to automatically test AI-based plant identification mobile apps. First, TestPlantID defines a set of test contexts to systematically depict the contextual factors of plant identification app usage scenarios. Next, MRs based on predefined test contexts are defined for plant identification apps. These test-context-based MRs are leveraged for revealing inconsistent behaviors of apps under test. Then, by applying image transformations and photographing real-world plants, TestPlantID generates images with different test contexts for performing MT. Finally, TestPlantID is leveraged to test three real-world plant identification mobile apps and successfully detect inconsistent results.

The key contributions of this paper are as follows:

- We present a systematic metamorphic testing approach to automatically test AI-based plant identification mobile apps under different test contexts. We leverage MRs based on test contexts for detecting inconsistent behaviors of apps.
- We perform a case study to indicate the feasibility and effectiveness of the proposed metamorphic testing approach on three real-world plant identification mobile apps. The results show that TestPlantID found more than five thousand inconsistent behaviors across three apps. Besides, we also investigate what extent do different test-context-based MRs reveal inconsistent behaviors.

The remainder of this paper is structured as follows: Sect. 2 gives a general summary of related work. In Sect. 3, the approach of this current study is elaborated. In Sect. 4, we present the case study and the study's validity. Section 5 shows the conclusion and future work.

2   Related Work

2.1   Metamorphic Testing

Metamorphic testing (MT) is a property-based software testing technique, which has been leveraged in many domains for addressing the test oracle problem and test case generation problem [7]. Since Chen et al. [8] introduced MT in 1998, MT has been an attractive research topic in software engineering. It has been used on debugging [9,10], fault localization [11], fault tolerance [12], and program repair [13]. Recently, MT has been proved to be an effective AI-based software testing approach. It has successfully helped detect a large number of real-life faults. Tian et al. [14] proposed a testing tool for Deep Neutral Network (DNN)-based autonomous driving systems called DeepTest. DeepTest has automatically detected potentially fatal behaviors in the system with MT. They simulated the actual driving scenes by applying a variety of image transformations and effect filters for transforming the original driving scene images. Zhang et al.

[15] proposed the first GAN-based MT approach to delivering driving scene-based test generation with various weather conditions for detecting inconsistent behaviors of autonomous driving systems. Zhou et al. [16] found fatal errors in the LiDAR Obstacle Perception system of the Baidu Apollo autonomous driving system by employing MT. In addition to the automatic driving system, MT has also been applied to ML classifiers [17,18], Google map App [19], search engines [20], facial age recognition software [3], and object detection system [21], etc., all of which have achieved good results.

The key element of MT is a set of effective MRs, which are necessary features of the target function or algorithm in relation to multiple inputs and their expected output. Some studies presented various approaches to systemically generate MRs [22–24]. Even though many MRs have been identified for various application domains [7], there is a lack of approaches identifying MRs from the test context perspective. In this paper, TestPlantID leverages an MT approach for plant identification apps, where MRs are defined based on predefined test contexts.

2.2 Testing and Verification of AI-Based Software Systems

Traditional software is implemented by developers with carefully designed specifications and programming logic. It is tested with test cases which are designed based on specific coverage criteria. For traditional software applications, testing is efficient and effective. However, the current practice of testing AI applications lags far behind the maturity of testing traditional software applications [25]. More and more work focused on testing ML-based software, including proposing new testing evaluation criteria, generation of test cases, etc. Pei et al. proposed DeepXplore [26], the first white-box testing framework for real-world Deep Neural Network (DNN) systems. They introduced neuron coverage as s systematic metric for measuring how much of the internal logic of a DNN has been tested. Ma et al. [27] extended the concept of neuron coverage. They proposed a set of multi-granularity test criteria called DeepGauge for the DNN system. Sun et al. [28] proposed four test coverage criteria that are tailored to the distinct features of DNN inspired by the MC/DC coverage criteria. Besides, testing techniques for traditional software have been recently applied for AI-based software systems, including fuzz testing [29–31], mutation testing [32,33], metamorphic testing [34,35], and also symbolic execution [36–38].

As a typical AI-based software, the image recognition system detects and recognizes objects in images by referring to a database of images. However, testing image recognition grave great challenges. Zhu et al. [25] proposed a new method called Datamorphic Testing, which consists of three components: a set of seed test cases, a set of datamorphisms for transforming test cases, and a set of metamorphisms for checking test results. They validated the proposed approach on four real industrial face recognition systems. Tao et al. [3] performed a case study on a realistic facial age recognition provided by Alibaba Company using MT. To the best of our knowledge, TestPlantID is the first work that focuses on testing AI-based plant identification mobile apps.

3   Approach

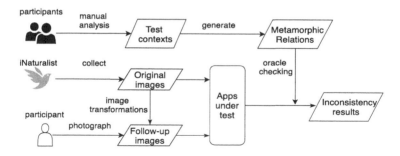

Fig. 2. The framework of TestPlantID

The details of TestPlantID are presented in this section. Figure 2 shows the framework of TestPlantID.

3.1   Test Context Construction

As a typical intelligent software artifact, AI-based plant identification mobile apps allow users to identify diverse plants simply by photographing them or uploading images with users' smartphones. It provides several possible identification results for the uploaded images instantly. This kind of AI-based software is developed based on advanced ML algorithms through large-scale plant images training. As is mentioned before, AI-based mobile apps usually involve contextual factors, such as scenario, location, time, and stakeholders. Many mobile apps with AI functions generated inconsistent behaviors for the same test input when context conditions are changed. Thereby, the relevant real-world contexts such as image rotation, translation, lighting, or the distance between plant and smartphone could be leveraged to test the robustness of AI-based plant identification apps.

It is worth noting that, we only consider the contextual factors relevant from usage scenarios of plant identification apps. In this paper, a test context refers to a major factor or characteristic of an environmental condition when testing AI-based apps. To construct a set of test contexts for the plant identification app, two participants are involved in this process. Both participants are postgraduate students majoring in software engineering and have more than 2 years of experience in conducting research on testing AI software. Inspired by the MRs proposed by the literature [14,39], each participant defines several real-world contexts for plant identification. If they select the same contextual factors, the context condition is considered as a test context. If there are disagreements, they discuss with each other to determine a final judgment. Finally, seven test contexts are determined for testing plant identification apps. Table 1 illustrates test contexts and definitions of plant identification apps.

Table 1. Test contexts and definitions of plant identification apps.

Test context	Definition
Lighting	Change the lighting condition when photographing the plant
Angle	Change the angle at which photograph the plant
Distance	Change the distance between the smartphone and the plant
Background	Change the background scenario of the plant
Position	Change the position of the plant in the image when photographing it
Rotation	The original image is rotated by a specific degree
Image clarity	Change the clarity of the plant image

3.2 Metamorphic Testing for Plant Identification Apps

One of the major challenges in testing AI-based plant identification apps lies in the lack of test oracle, which is also known as "oracle problem". To avoid this issue, TestPlantID adopts MT to test plant identification apps under different test contexts.

The key insight of MT is that even though it is hard to specify the correct behavior of the AI-based plant identification, we can define the relationships between the results of the original image and the corresponding follow-up image. TestPlantID leverages test-context-based MRs for detecting inconsistent behaviors of apps. For example, for the same plant, the output identification results should not change under different angle context conditions, such as 90-degree and 45-degree.

Formally, for an AI-based plant identification app PlantID, given an original image x, X is the database of original images to be identified. TestPlantID defines test context transformations T that simply change the context conditions of plant identification usage scenarios. Let $\tau(x)$ be a follow-up image which is generated by applying a test context transformation t on x. $PlantID(x)$ is the top-k identification results of image x. $PlantID(\tau(x))$ is the top-k identification results of the follow-up image $\tau(x)$. Then, MRs based on test contexts can be defined as follows:

$$\forall x \in X, \forall \tau \in T, PlantID(x) = PlantID(\tau(x)) \tag{1}$$

In this case, test context transformations T can simply change context conditions without impacting the identification results for each plant. One MR is defined that $PlantID(x)$ and $PlantID(\tau(x))$ should be identical when test context t is changed. If $PlantID(x)$ and $PlantID(\tau(x))$ are of significant difference, we can conclude that the plant identification app has wrongly behaved under a specific test context.

3.3 Follow-Up Image Generation

To perform MT on plant identification apps, we need to generate follow-up plant images under different test contexts. The goal of this part is to generate follow-up images under different test context conditions.

In recent studies, DeepTest performs simple image transformations and effect filters on original images to mimic real-world road scenes [14]. In this paper, five different types of simple image transformations (changing brightness, blurring, translation, cropping, rotation) are leveraged to automatically generate follow-up images, in order to implement predefined test-context-based MRs. Changing brightness is a linear transformation, which is performed by adding or subtracting a constant parameter β to each pixel's current value [14]. Cropping transformation is used to select an area of specified size on the original image. Translation and rotation are affine transformations [40], which are the linear mapping between two images that preserve points, straight lines, and planes. It is worth noting that we use crop transformation to mimic the different distances between the smartphone and the plant. By applying blurring effects on original images, the condition of poor image clarity due to camera lens distortions or subjective factors of the photographer can be implemented. Translation transformation is leveraged to simulate the different positions of the plant in the image.

We use OpenCV [41] to implement the brightness, cropping, translation, rotation, and blur image transformations. Each transformation has six parameters. The transformations and corresponding parameters are shown in Table 2.

Table 2. Image transformations and parameters for generating follow-up images.

MR	Image transformation	Parameters	Parameter ranges
MR-lighting	Brightness	β	$(-60, 60)$ step 20
MR-distance	Cropping	$(y \cdot \frac{n}{32} : y \cdot \frac{32-n}{32}, x \cdot \frac{n}{32} : x \cdot \frac{32-n}{32})$	n from 1 to 6
MR-position	Translation	(tx, ty)	$(60, 60)$ to $(110, 110)$
			step $(10, 10)$
MR-rotation	Rotation	q (degree)	$(30, 80)$ step 10
MR-image clarity	Blur averaging	Kernel size	5*5, 6*6
	Blur Gaussian	Kernel size	7*7
	Blur Median	Aperture linear size	3,5
	Blur Bilateral	Diameter, sigmaColor, sigmaSpace	9, 75, 75

For MR-background, Python library removebg [42] is leveraged for removing the background of the original image and keep the plant object. Then, we photographed six different images, which are used as new background images. The plant object is inserted into these six background images. Figure 3 illustrates an

Fig. 3. Two images with different background of a plant

original plant image and the corresponding follow-up image after transforming the background.

For MR-angle, one participant was involved in photographing various plants from seven different angles. We use the degree of the angle from the smartphone plane to the plant object to define the seven different angles. The seven different angles of a plant are defined as follows: 45-degree, 75-degree, 180-degree, and four different shooting sides of 90-degree. As is illustrated in Fig. 1, they are two different angles of the same plant photographed by the participant. In the experiment, we use the images photographed from the 45-degree angle as the original image.

4   Case Study

In this section, we perform a case study to indicate the feasibility and effectiveness of the proposed MT approach to plant identification apps under different test contexts. To evaluate the ability of the proposed testing approach, we investigate whether different MRs based on test contexts could trigger inconsistent behaviors of plant identification apps.

4.1   Dataset

We use the dataset from the iNaturalist 2018 Competition [43], which is a part of the FGVC5 workshop at CVPR. iNaturalist is an object identification app, which can identify plants and animals. There are a total of 2917 plant species in the dataset, with 118800 training and 8751 validation images. We selected randomly 200 plant images from validation images as original images. By applying image transformations on original images and photographing real-world plants, a total of 8400 follow-up images were generated.

4.2   Plant Identification Apps Under Test

We have selected three AI-based plant identification mobile apps from Google Play Store as subjects. They are developed by advanced AI algorithms and trained by large-scale plant images.

- PlantSnap [44]: PlantSnap has an average rating of 3.7 in the Google Play Store. It has a database of more than 625000 plants, flowers, and mushrooms. PlantSnap has been widely used over 35 million plant lovers in more than 200 countries. It provides a maximum of 10 possible identification results for each plant image.
- PlantNet [45]: PlantNet has an average rating of 4.6 in the Google Play Store. It is organized in different databases, such as world flora, weeds, useful plants of tropical Africa, etc. In this paper, we choose the world flora database for testing. There are a total of more than 22000 species, with more than 260000 images in the world flora dataset. PlantNet outputs the possible results together with corresponding probabilities.

– PictureThis [46]: PictureThis has an average rating of 4.2 in the Google Play Store. It claimed that it is capable of identifying more than 10000 plant species with an accuracy of 98%, even better than most human experts. PictureThis returns 3 possible results for each plant image.

4.3   Evaluation Metrics

Metrics of Inconsistent Behaviors of Plant Identification Apps: In this paper, we need to evaluate whether different MRs based on test contexts could trigger inconsistent behaviors of plant identification apps. The original images and follow-up images are used to test apps. If the results of the original image and follow-up image violate MRs, it means the app has inconsistent behaviors under specific test contexts. In the case study, we select top-1 and top-3 results of each test image.

For the top-1 result, we compute the number of inconsistent behaviors of original images. For the top-3 results, we compute the dissimilarity of results between the original image and the follow-up image. Given an original image x, its corresponding follow-up image f is generated by applying a test context transformation t. Let $Rx = PlantID(x)$ be the output of original image, $Rf = PlantID(\tau(x))$ be the output of follow-up image. To compute the inconsistency between Rx and Rf, we adopt the Jaccard distance to measure their dissimilarity. It has been used in measuring the inconsistent behaviors of AI software [39]. It is worth noting that we do not consider the order of identification results. The dissimilarity is defined as follows:

$$RD = 1 - \frac{|Rx \bigcap Rf|}{|Rx \bigcup Rf|} \tag{2}$$

Obviously, the higher the dissimilarity is, the lower the similarity between the original image result and the follow-up image result. Therefore, the RD value depicts the inconsistency of results between original and follow-up images.

4.4   Case Study Results

Effective MRs are the MRs with a higher chance of revealing failures, which are the key to perform MT. First, we check whether MRs based on test contexts could reveal inconsistent behaviors of the top-1 results between the original and follow-up images. Here, we focus on the number of follow-up images whose output violates MRs. For top-1 results, we compute the number of inconsistent behavior of three apps under different MRs.

Table 3 presents the number of inconsistent behavior of top-1 results across three apps under different MRs. From the table, we can observe that TestPlantID detects 2308, 2605, 1040 inconsistent behaviors of top-1 results for PlantNet, PlantSnap, and PictureThis respectively under all MRs. In total, 5953 inconsistent behaviors are found across all three apps. From Table 3, we can observe that the number of inconsistent behavior of PictureThis is the lowest under all

Table 3. Number of inconsistent behavior of three apps under different MRs

MR based on test context	PlantNet	PlantSnap	PictureThis
MR-angle	335	329	95
MR-background	419	668	159
MR-lighting	367	437	158
MR-image clarity	320	299	163
MR-rotation	218	318	148
MR-distance	325	165	151
MR-position	324	389	166
Total	2308	2605	1040

MRs. PlantSnap shows the worst robustness across three apps since it has the maximum number of inconsistent behaviors. Interestingly, some apps are more prone to inconsistent behaviors for some specific MRs than others. For example, PlantNet produces 325 inconsistent behaviors under MR-distance, while the other two apps produce half of that number. Similarly, we detect 668 inconsistent behaviors of PlantSnap with MR-background, but only 419 and 159 for Plant-Net and PictureThis respectively. Besides, PictureThis has the lowest number of inconsistent behavior under MR-angle. We can see from Table 3 that it is feasible and effective to automatically detect inconsistent behaviors of three apps under test by using test-context-based MRs we proposed.

To evaluate whether different test-context-based MRs could trigger inconsistent behaviors of top-3 results, we compute the dissimilarity of top-3 results between the original image and follow-up image. The RD value is computed

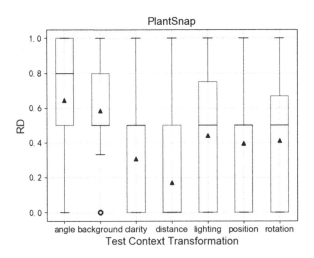

Fig. 4. RD distribution of each MR on PlantSnap

for each pair of test inputs (the original image with its corresponding follow-up image under a specific test context). Figures 4, 5 and 6 show the RD distribution under different MRs for PlantSnap, PlantNet, PictureThis respectively. In these figures, the mean RD value is depicted with a triangle label, and the median RD value is depicted with a solid line. We can observe that PictureThis performs more consistent than the other two apps under almost all seven MRs, with relatively lower median RD values and mean RD values.

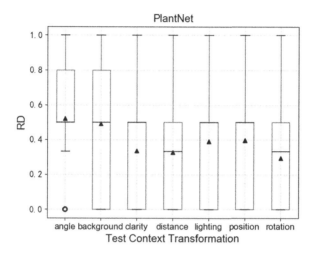

Fig. 5. RD distribution of each MR on PlantNet

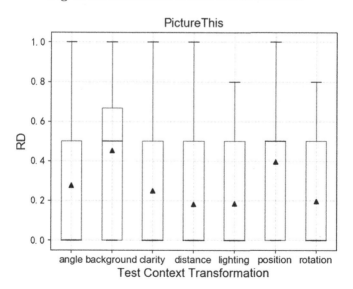

Fig. 6. RD distribution of each MR on PictureThis

MR-lighting, MR-rotation, and MR-distance have the relatively lower capability of revealing inconsistent behaviors on PictureThis. These MRs all have relatively lower median RD values equal to 0 and mean RD values around 0.2. It indicates that PictureThis could stay robust under different lighting, rotation, and distance test contexts. The possible reason is that those image transformations we leveraged to implement MR-lighting, MR-rotation and MR-distance could keep plant features of images (such as fruits, leaves, thorns, buds, or hair on the stem, which are the most characteristic organs), resulting in accurate identification results. Moreover, compared with other MRs, MR-background has the highest median and mean RD value on PictureThis, which indicates that PictureThis has the worst robustness to the background change. Moreover, MR-angle has a high capability of detecting inconsistent behaviors on PlantSnap and PlantNet, with median RD values up to 0.8 and 0.5 respectively. Compared with PlantSnap and PictureThis, MR-distance shows effectiveness in detecting inconsistent behaviors on PlantNet, with a median RD value and a mean RD value all close to 0.4. For PlantNet, MR-rotation has the lowest median RD value and mean RD value compared with other MRs. We also notice that all three apps behave similarly under the position test context condition. They all have a median RD value of 0.5 and a mean RD value of 0.4. MR-lighting and MR-position show similar performance at revealing inconsistent behaviors on PlantSnap and PlantNet. The median RD values are around 0.5 and mean RD values are around 0.4.

To sum up, different MRs based on test contexts not only effectively detects inconsistent behaviors of plant identification functions, but also potentially be useful for the measurement of the robustness of different AI-based apps under diverse context conditions.

4.5 Threats to Validity

In this paper, we generate follow-up plant images by applying simple image transformations and photographing real-world plants. However, even though we have carefully configured the parameters of image transformations, these follow-up images are not sufficient enough to cover all usage scenarios in the real-world. This could affect the chance of revealing inconsistent behaviors of apps. Besides, simple image transformations like changing the background of plants tend to be realistic, while they cannot sophisticatedly synthesize images with different complex usage scenarios. For example, the app user might photograph the plant from a moving car where is a good distance away from the plant. This kind of context scenario cannot be generated by simple image transformations. As the image processing techniques such as GAN become more and more advanced, we do expect that the generated images could be more close to real usage scenarios. Furthermore, We exploited the test contexts defined by two participants with the experience of AI software testing and good English knowledge. In this case, the final determination of test contexts was discussed by two participants, we still cannot avoid subjective factors affecting the construction of test contexts.

Moreover, we validated our approach on a dataset from iNaturalist competition with 200 original plant images. The dataset is relatively small. The limited number of test data could also be a threat to validity. Future work will conduct a large-scale empirical study to address this threat.

5   Conclusion and Future Work

In this paper, we proposed and validated TestPlantID, a metamorphic testing approach to automatically test AI-based plant identification mobile apps under different test contexts. To evaluate the robustness of AI-based plant identification apps, we leverage MRs based on test contexts to detect inconsistent behaviors. By applying image transformations and photographing real-world plants, follow-up images are generated for performing MT. Furthermore, a case study on three plant identification mobile apps is performed to indicate the feasibility and effectiveness of the proposed testing approach.

For future work, we will evaluate TestPlantID on more plant identification mobile apps. A large-scale empirical study with more datasets will be conducted. Meanwhile, we plan to implement an automatic testing tool for detecting inconsistent behaviors.

References

1. Zhang, J., Harman, M., Ma, L., Liu, Y.: Machine learning testing: survey, landscapes and horizons. arXiv preprint arXiv:1906.10742 (2019)
2. Amershi, S., et al.: Software engineering for machine learning: a case study. In: Proceedings of the 41st International Conference on Software Engineering: Software Engineering in Practice (ICSE-SEIP), Montreal, QC, Canada, pp. 291–300 (2019)
3. Tao, C., Gao, J., Wang, T.: Testing and quality validation for AI software-perspectives, issues, and practices. IEEE Access **7**, 120164–120175 (2019)
4. Yin, Y., Chen, L., Xu, Y., Wan, J.: Location-aware service recommendation with enhanced probabilistic matrix factorization. IEEE Access **6**, 62815–62825 (2018)
5. Gao, J., Tao, C., Jie, D., Lu, S.: Invited paper: what is AI software testing? and why. In: IEEE International Conference on Service-Oriented System Engineering (SOSE), San Francisco East Bay, CA, USA, pp. 27–2709 (2019)
6. Barr, E., Harman, M., McMinn, P., Shahbaz, M., Yoo, S.: The oracle problem in software testing: a survey. IEEE Trans. Softw. Eng. **41**(5), 507–525 (2015)
7. Chen, T.Y., et al.: Metamorphic testing: a review of challenges and opportunities. ACM Comput. Surv. **51**(1), 4:1–4:27 (2018)
8. Chen, T.Y., Cheung, S., Yiu, S.: Metamorphic testing: a new approach for generating next test cases. Technical report HKUST-CS98-01. Department of Computer Science, Hong Kong University of Science and Technology, Hong Kong (1998)
9. Chen, T.Y., Tse, T., Zhou, Z.: Semi-proving: an integrated method for program proving, testing, and debugging. IEEE Trans. Softw. Eng. **37**(1), 109–125 (2011)
10. Jin, H., Jiang, Y., Liu, N., Xu, C., Ma, X., Lu, J.: Concolic metamorphic debugging. In: Proceedings of the IEEE 39th Annual Computer Software and Applications Conference (COMPSAC), Los Alamitos, CA, pp. 232–241 (2015)

11. Xie, X., Wong, W.E., Chen, T.Y., Xu, B.: Spectrum-based fault localization: testing oracles are no longer mandatory. In: Proceedings of the 11th International Conference on Quality Software (QSIC), Los Alamitos, CA, pp. 1–10 (2011)

12. Liu, H., Yusuf, I.I., Schmidt, H.W., Chen, T.Y.: Metamorphic fault tolerance: an automated and systematic methodology for fault tolerance in the absence of test oracle. In: Companion Proceedings of the 36th International Conference on Software Engineering (ICSE Companion), New York, NY, pp. 420–423 (2014)

13. Jiang, M., Chen, T.Y., Kuo, F.C., Towey, D., Ding, Z.: A metamorphic testing approach for supporting program repair without the need for a test oracle. J. Syst. Softw. **126**, 127–140 (2017)

14. Tian, Y., Pei, K., Jana, S., Ray, B.: DeepTest: automated testing of deep-neural-network-driven autonomous cars. In: Proceedings of the 40th International Conference on Software Engineering (ICSE), Gothenburg, Sweden, pp. 303–314 (2018)

15. Zhang, M., Zhang, Y., Zhang, L., Liu, C., Khurshid, S.: DeepRoad: GAN-based metamorphic testing and input validation framework for autonomous driving systems. In: Proceedings of the 33rd IEEE/ACM International Conference on Automated Software Engineering (ASE), Montpellier, France, pp. 132–142 (2018)

16. Zhou, Z., Sun, L.: Metamorphic testing of driverless cars. Commun. ACM **62**(3), 61–67 (2019)

17. Murphy, C., Kaiser, G.E., Hu, L., Wu, L.: Properties of machine learning applications for use in metamorphic testing. In: Proceedings of the 20th International Conference on Software Engineering and Knowledge Engineering (SEKE), San Francisco, CA, USA, pp. 867–872 (2008)

18. Xie, X., Ho, J.W., Murphy, C., Kaiser, G., Xu, B., Chen, T.Y.: Testing and validating machine learning classifiers by metamorphic testing. J. Syst. Softw. **84**(4), 544–558 (2011)

19. Brown, J., Zhou, Z. Chow, Y.: Metamorphic testing of navigation software: a pilot study with google maps. In: 51st Hawaii International Conference on System Sciences (HICSS), Hilton Waikoloa Village, Hawaii, USA, pp. 1–10 (2018)

20. Zhou, Z., Xiang, S., Chen, T.Y.: Metamorphic testing for software quality assessment: a study of search engines. IEEE Trans. Softw. Eng. **42**(3), 264–284 (2016)

21. Wang, S., Su, Z.: Metamorphic testing for object detection systems. arXiv preprint arXiv:1912.12162 (2019)

22. Chen, T.Y., Poon, P., Xie, X.: METRIC: METamorphic Relation Identification based on the Category-choice framework. J. Syst. Softw. **116**, 177–190 (2016)

23. Zhang, J., et al.: Search-based inference of polynomial metamorphic relations. In: Proceedings of the 29th ACM/IEEE International Conference on Automated Software Engineering (ASE), New York, pp. 701–712 (2014)

24. Zhu, H.: A tool for automated Java unit testing based on data mutation and metamorphic testing methods. In: Proceedings of the 2nd International Conference on Trustworthy Systems and Their Applications (TSA), Los Alamitos, CA, pp. 8–15 (2015)

25. Zhu, H., Liu, D., Bayley, I., Harrison, R., Cuzzolin, F.: Datamorphic testing: a method for testing intelligent applications. In: IEEE International Conference On Artificial Intelligence Testing (AITest), Newark, CA, USA, pp. 149–156 (2019)

26. Pei, K., Cao, Y., Yang, J., Jana, S.: Deepxplore: Automated whitebox testing of deep learning systems. In: Proceedings of the 26th Symposium on Operating Systems Principles (SOSP), pp. 1–18. Shanghai, China (2017)

27. Ma, L., et al.: DeepGauge: multi-granularity testing criteria for deep learning systems. In: Proceedings of the 33rd ACM/IEEE International Conference on Automated Software Engineering (ASE), Montpellier, France, pp. 120–131 (2018)

28. Sun, Y., Huang, X., Kroening, D.: Testing deep neural networks. arXiv preprint arXiv:1803.04792 (2019)

29. Guo, J., Jiang, Y. Zhao, Y. Chen, Q. Sun, J.: DLFuzz: differential fuzzing testing of deep learning systems. In: Proceedings of the 2018 26th ACM Joint Meeting on European Software Engineering Conference and Symposium on the Foundations of Software Engineering (ESEC/SIGSOFT), Lake Buena Vista, FL, USA, pp. 739–743 (2018)

30. Odena, A., Olsson, C., Andersen, D., Goodfellow, I.: TensorFuzz: debugging neural networks with coverage-guided fuzzing. In: Proceedings of the 36th International Conference on Machine Learning (ICML), Long Beach, California, USA, pp. 4901–4911 (2019)

31. Xie, X., Chen, H., Li, Y., Ma, L., Liu, Y., Zhao, J.: Coverage-guided fuzzing for feedforward neural networks. In: 34th IEEE/ACM International Conference on Automated Software Engineering (ASE), San Diego, CA, USA, pp. 1162–1165 (2019)

32. Ma, L., et al.: DeepMutation: mutation testing of deep learning systems. In: Proceedings of the 29th IEEE International Symposium on Software Reliability Engineering (ISSRE), Memphis, TN, pp. 100–111 (2018)

33. Shen, W., Wan, J., Chen, Z.: MuNN: mutation analysis of neural networks. In: IEEE International Conference on Software Quality, Reliability and Security Companion (QRS-C), Lisbon, pp. 108–115 (2018)

34. Ding, J., Kang, X., Hu, X.: Validating a deep learning framework by metamorphic testing. In: 2017 IEEE/ACM 2nd International Workshop on Metamorphic Testing (MET), Buenos Aires, pp. 28–34 (2017)

35. Murphy, C., Shen, K., Kaiser, G.: Using JML runtime assertion checking to automate metamorphic testing in applications without test oracles. In: International Conference on Software Testing Verification and Validation (ICST), Denver, Colorado, USA, pp. 436–445 (2009)

36. Sun, Y., Wu, M., Ruan, W., Huang, X., Kwiatkowska, M., Kroening, D.: Concolic testing for deep neural networks. In: 33rd IEEE/ACM International Conference on Automated Software Engineering (ASE), Montpellier, France, pp. 109–119 (2018)

37. Gopinath, D., Wang, K., Zhang, M., Pasareanu, C., Khurshid, S.: Symbolic execution for deep neural networks. arXiv preprint arXiv:1807.10439 (2018)

38. Gopinath, D., Zhang, M., Wang, K., Kadron, B., Pasareanu, C., Khurshid, S.: Symbolic execution for importance analysis and adversarial generation in neural networks. In: IEEE 30th International Symposium on Software Reliability Engineering (ISSRE), Berlin, Germany, pp. 313–322 (2019)

39. Zhang, Z., Xie, X.: On the investigation of essential diversities for deep learning testing criteria. In: IEEE 19th International Conference on Software Quality, Reliability and Security (QRS), Sofia, Bulgaria, pp. 394-405 (2019)

40. Affine Transformation (2015). https://www.mathworks.com/discovery/affine-transformation.html

41. Open Source Computer Vision Library (2015). https://github.com/itseez/opencv

42. Removebg (2019). https://github.com/remove-bg

43. iNaturalist 2018 Competition (2018). https://github.com/visipedia/inat_comp/tree/master/2018

44. PlantSnap. https://play.google.com/store/apps/details?id=com.fws.plantsnap2

45. PlantNet. https://play.google.com/store/apps/details?id=org.plantnet

46. PictureThis. https://play.google.com/store/apps/details?id=cn.danatech.xingseus

Author Index

Printed in the United States
By Bookmasters